DUMPED CHEMICAL MUNITIONS IN THE BALTIC AND SKAGERAK;
Legal obligations and technical options to reduce incidents

Egbert K. Duursma, Jam G. de Vries and Jorri C. Duursma (eds)

Revised and updated original document[1]
Dumped Chemical Weapons in the Sea; Options

Authors of actual and first document are cited in 10.5.

---

[1] Dumped Chemical Weapons in the Sea; options, E.K. Duursma (ed.) Published through the Dr. Alfred H. Heineken Foundation for the environment. 1999

KEYWORDS:
Chemical weapons
Baltic Sea and Skagerak
Mustard gas
Dumping
Sarcophaging
Micro-waves techniques
First-aid
Insurance
Legal responsibility of States
Helsinki Commission
UNCLOS III

LIST OF CONTENTS

PREFACE

In 1999 questions[2] to the Dutch government were posed by the CDA fraction in the Second Chamber of the Dutch Parliament concerning the risks of dumped chemical munitions in the Baltic Sea and the Skagerak **(Fig. 1)**, which is part of the North Sea. The Minister replied: *[Since 1990, the Netherlands has a regulation on the Assistance for Dutch fishermen to be under certain conditions eligible for compensation if they fish explosives in their nets. For now I see no need for the introduction of additional arrangements concerning possible injury and other damages resulting from calamities with chemical weapons considering the extremely low chances of such calamities. In addition, for the specific problems in the Baltic Sea Region, the Helsinki Commission (HELCOM) is the most appropriate forum for specific measures to be determined.]*

Since then, Helcom[3] has intensified its reporting, which resulted in their document Chemical munitions dumped in the Baltic Sea, Report 142, which demonstrates that since 1999 a great number of incidents have been observed in the Baltic Sea either with fishermen and with beach visitors, in which 5000 kg of retrieved mustard gas was involved. Knowing that 5 gram mustard gas is lethal, the chances of calamities are not that low as suggested in the reply of the Dutch Government.

The major problem is that lumps of this mustard gas increasingly leak from corroded dumped munitions. Mustard gas can only hydrolyse in sea water when it leaks as molecules. The lumps may remain at the sea bottom for years. The corrosion of bombs, canisters, drums and grenades is steadily progressing.[4] Moreover, any criminal could sail to a dump site and explode a shipwreck, or threaten to do so for terroristic or other reasons.

The fact that the Helcom Commision still concludes that nothing can be done, since it is too dangerous to recover chemical munitions, does not exclude the border Baltic and North Sea States to have a **legal** obligation towards the problem and seek for appropriate solutions.

Maps and pilot studies should be made on the bottom currents of the involved seas and marine engineers can at least carry out tests in order to find reasonable solutions to reduce incidents. These options should seriously be envisaged.

Fortunately a consortium of Baltic institutes like CHEMSEA[5] has taken initiatives in this direction. Hopefully States will joint with sponsoring the needed pilot studies and perhaps emergency technical actions as proposed in this document and apply to which they are legally obliged.

---

[2] Tweede Kamer der Staten-Generaal, Vragen gesteld door de leden der Kamer, met daarop door de regering gegeven antwoorden, No. 112, 13 October 1999, page 235/

[3] HELCOM, 2013. Chemical Munitions Dumped in the Baltic Sea. Report of the *ad hoc* Expert Group to Update and Review the Existing Information on Dumped Chemical Munitions in the Baltic Sea (HELCOM MUNI) BSEP) No. 142 Baltic Sea Environment Proceeding pages 128.

[4] Malyshev (1996), p. 93.

[5] CHEMSEA, Results from the CHEMSEA project –Chemical munitions Search and Assessment. ISBN:978-83-936609-1-9.

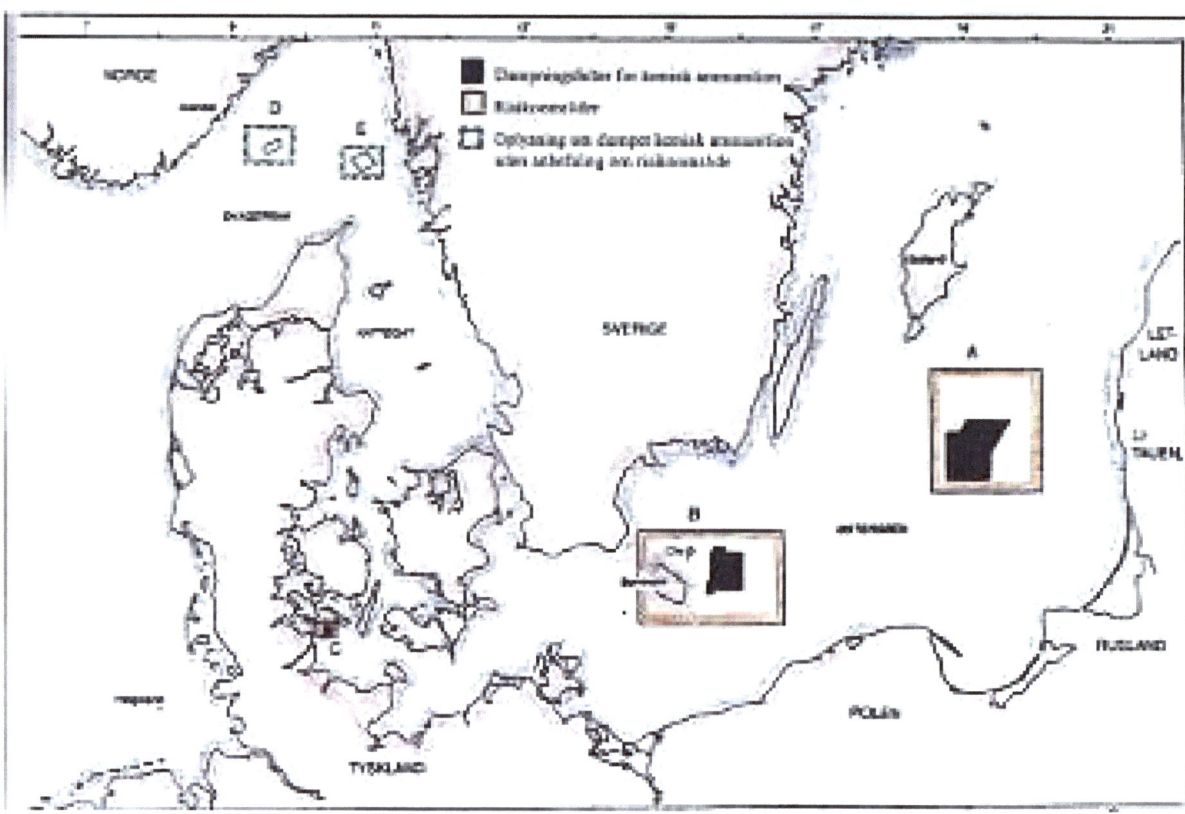

Fig. 1. Baltic Sea and Skagerak containing the Helsinki Commission Risk Areas where recovery of CW munitions is not recommended.[6]. Map obtained from Mr. Kjeld F. Jørgensen, Miljø og Energiministeriet, København, Denmark, lead country for the Helsinki Commission concerning dumped chemical weapons.
A: 55°50'-56°40'N & 18°30'-20°00'E; **B**: 54°50'-55°30'N & 14°30'-16°30'E;
C: 54°45'-54°52'N & 10°00'-10°20'E; **D**: 58°10'-58°25'N & 09°10'-09°50'E;
E: 58°07'N & 10°47'E[7]

---

[6]HELCOM CHEMU (1995) (Final report to Helsinki Commission).

[7]These coordinates are not indicated on the official nautical maps (see Annex IV).

CONCLUSIONS AND STEPS TO BE TAKEN
E.K.Duursma and B.T.Surikov

CONCLUSIONS
Baltic Sea
* Although leaked chemical warfare (CW) gases rapidly hydrolyse in sea water, potential risks of serious contamination will exist for many decades for sailors, fishermen and coastal visitors of the Baltic Sea, in particular concerning contact with lumps of mustard gas.
* Since 72% of the CW agents is contained in aircraft bombs, which are already in various stages of corrosion and 63% of all CW agents is S-mustard gas, the loss by lumps of mustard, spread over the seabed in the neighbourhood of dump sites, represents the greatest danger.
* There remain sites at risk outside the Allied Forces dump sites, since during the disposal procedures before 1948, ammunition was thrown overboard between the discharge ports of Wolfgast and Peenemünde and the dump sites east of Bornholm and south of Gotland.
* Blue prints of emergency plans should be available in order to confine, bury or destroy ammunition at risk.
* For that reason bottom current maps should be available to predict the transport of lumps over the sea bottom.
* Test should be made in hydraulic laboratories about the correlation water current and lump transport.
* Modelling may predict the hotspots in the future.

Skagerak
* Most dumped CW ammunition is contained in deliberately sunken vessels.
* Implosion of the chemical weapon's cargo in these vessels, due to its own weight, may increase loss of CW agents, in particular from corroded aircraft bombs.
* Several wrecked ships have been located outside the dump site as indicated on the nautical charts.
* Steps should be taken to sarcophage the wrecked ships in order to avoid an eventual contamination of the eastern North Sea and coastal waters of Norway, Sweden and Denmark.
* Similar studies and modelling as for the Baltic Sea should be made for the Skagerak and adjacent north Sea, in particular where rest streams pass along the Norwegian coast.

General
* Although the Helsinki Commission, through its *ad hoc* Working Group HELCOM CHEMU has correctly evaluated the present risks of the dumped CW ammunition, it is obvious that the authorities of the Baltic States and Norway should be prepared to act jointly in **emergency** situations. They arc bound, both by the Helsinki Convention and UNCLOS-III (Third United Nations Convention on the Law of the Sea) to take "appropriate measures".

* The Helcom No. 142 report[8] supplies at present excellent guidelines to authorities in dealing with chemical munitions caught by fishermen, as prepared by HELCOM CHEMU and accepted by the Helsinki Commission.
* In a number of cases, confinement and destruction measures can be taken to prevent emergency situations. They are:
    * Sarcophaging of the wrecked ships, located in the Baltic Sea and Skagerak.
    * Seabed burial or/and destruction of scattered CW ammunition which are not within the Allied Forces dump sites east of Bornholm and south of Gotland.
* Pilot tests should be made (which can be held at low costs), to prepare the best method of burial munitions into the sediment.
* Pilot tests should be made to determine the transport velocities of mustard gas lumps in relation to bottom current speeds. From these modelling can be tried to determine the distribution of lumps over large areas.

1. INTRODUCTION
E.K.Duursma and B.T.Surikov

On 24 April 1994 the Polish captain of the fishing vessel LEB-5 tried to deposit in Nexø, Bornholm, a bomb containing warfare gas which he had captured in his nets. He was fined by the local court for polluting the environment and had to pay for disarming the bomb or have his vessel confiscated.[9]

In the summer of 1998 the Russian Research vessel "Professor Shtockman" dredged soil samples from the Baltic Sea containing a large black-brown substance, smelling like fresh hay. The scientists on board suspected this substance to be dangerous. They put on gas masks and placed the material in a hermetically closed container. It was a lump of viscous mustard gas, sufficient to kill the whole crew of the vessel.[10]

These two examples characterise the problems posed by World War II German chemical warfare (CW) munitions dumped by the Allied Forces in the Baltic Sea and the Skagerak **(Fig. 2)**. The Allied Forces and the defeated German nation did not have the technology for safely neutralising these chemical weapons on land and at the 1945 Potsdam conference[11] the leaders of the USSR, USA, UK and France decided to dump these munitions deep in the Atlantic. Unfortunately this was neither technically nor financially possible.

The objectives of this synopsis are to present:
* Legal responsibilities of States.
* Options of confinement of wrecked ships and their CW cargo and separately dumped CW munitions.
* Options on the burial of dumped munitions at risk.
* Health aspects concerning retrieved CW agents.
* Insurance of sailors and coastal visitors endangered by CW agents.

---

[8] HELCOM, 2013 Chemical Munitions Dumped in the Baltic Sea. Report of the *ad hoc* Expert Group to Update and Review the Existing Information on Dumped Chemical Munitions in the Baltic Sea (HELCOM MUNI) Baltic Sea Environment Proceeding (BSEP) No. 142 Number of pages: 128

[9] Korzeniewski (1994).

[10] The location of this incident could not be obtained before the date (May 1999) this synopsis was going to press. The incident has not yet been reported to the Lead Country (Denmark) on dumped chemical munitions (closing date June 1999).

[11] Art. 3 of the Potsdam agreement,(...).*all war material should be distributed to the Allies or destroyed (...)*, cited by Stock (1996), p. 50.

Fig. 2. Photographs of a liberty ship, loaded with CW ammunition being sunk after WW-II. Photographs reproduced from the video-film, ⬚Cain's Smoke⬚. Permission obtained from Creative Ass. Film Program - the XXth Century, Moscow.

2. HISTORY
B.T.Surikov

In spite of the fact that the Wehrmacht poisoned about 10,000 Soviet servicemen and civilians in May 1942 in the stone quarries of Adzhimushkai near the town of Kerch at the Crimean peninsula and the application of Cyclone-B (hydrocyanic acid) in concentrations camps, Nazi Germany did not use its arsenal of chemical weapons in WW-II. However, its Wehrmacht Chemical Corps was ready for mass application of toxic agents and had at their disposal bombs, shell, mines of different calibre's as well as fougasses, hand grenades and poisonous smoke-pots. They were well equipped with special trucks to contaminate the terrain rapidly.

Fig. 3. Example of an early morning attack - see long shadows of soldiers on the right side - by the German Army during WW-I with suffocating warfare gas on the eastern front. The picture was taken by a Russian pilot and first published by l'Iskra, a supplement of the illustrated journal Rousskoie Slovo on 1 November 1915 under the title *Cain's Smoke*, photograph belonging to the collection of N.Verkhovaky. Reproduced from L'Illustration (1985) Nr. 11, Années 1914-1915, p. 243. Few pictures exist of the horrors of Ypres, Belgium, where the German Army on 17 July 1917 applied mustard gas.[12] A total of 125,000 tons CW agents were used in the battle-field during WW-I, heavily injuring 1,300,000 men, most of them died thereafter (information from the Military Encyclopaedic Dictionary of the former USSR Ministry of Defence). **Right below:** Major General Boris T. Surikov, initiator of the video film of the same name, *Cain's Smoke*, ⬜Creative Ass. Film Program - the XXth Century⬜. Address see Annex VI, Acknowledgements.

The Soviet military intelligence disclosed in time Hitler's intention to use chemical warfare agents at the eastern front in the summer of 1942. Through diplomatic channels of neutral states Marshal Stalin warned Hitler that the Soviet Union would undertake retaliatory steps if Hitler used chemical weapons in combat actions.

On June 5th 1942, President Roosevelt on behalf of the USA and Great Britain also warned Hitler that in response to chemical hostilities against any of the Allied States, these countries would do the same. Already at the beginning of 1942 the USA and Great Britain arranged the production of rather efficient organophosphorus toxic agents. These warnings served Hitler as a cold shower, and chemical weapons were not used in mass quantities during WW-II. (See also **Figs 3-5**)

---

[12]MEDEA (1997).

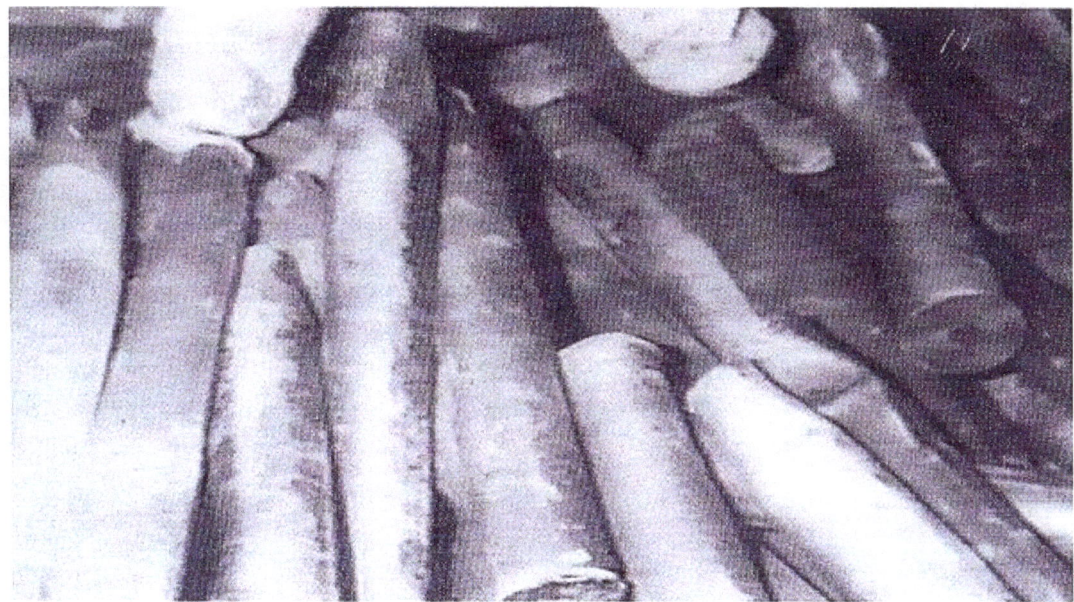

Fig. 4. CW artillery shells as discovered by the Allied Forces on German territory after WW-II.

Fig. 5. CW aircraft bomb as photographed on the seabed in the Baltic Sea. Photographs are reproduced from the Video film Cain's Smoke. Permission obtained from Creative Ass. Film Program - the XXth Century, Moscow.

3. DUMPED CW AGENTS IN EUROPEAN SEAS
B.T.Surikov and E.K.Duursma

3.1. Amounts of German CW agents discovered after WW-II

Before and during World War II, the German warfare chemical industry produced and accumulated 65,000 tons (net weight) of mustard gas, tabun, chloro-acetophenone and different arsenic containing compounds **(Fig. 6)**,[13] of which mustard gas was the major compound.

These were discovered after World War II by the Allied Forces and amounted to 290 - 300 thousand tons gross weight,[14] distributed over the occupation zones of the USA, UK, France and USSR **(Fig. 7)**, of which about 80% has been disposed at sea.

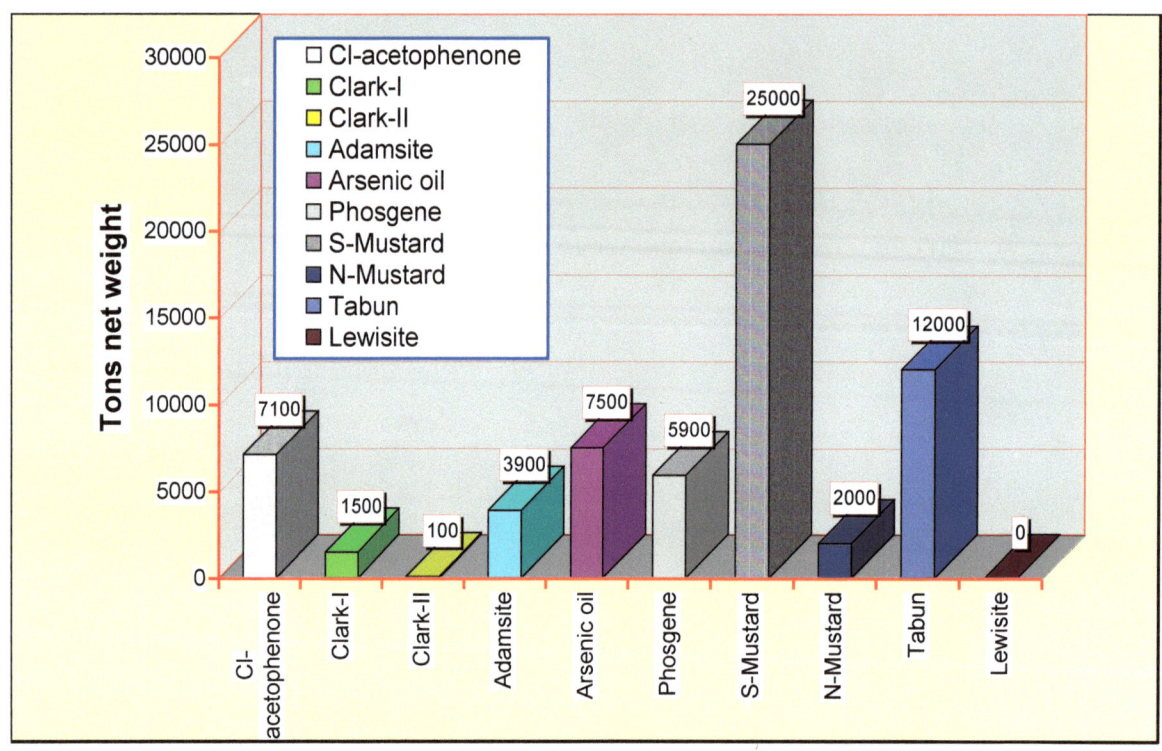

Fig. 6. Stocks of German chemical warfare (CW) agents produced and accumulated before and during WW-II.

3.2. Kind of ammunition and CW agents

Up to 600,000 units were counted by the Soviets in their occupation zone; the majority of these units were artillery projectiles[15] **(Fig 8)**. However by weight, the aircraft bombs contained 72% of all CW agents[16] **(Fig. 9)**. These munitions mainly contained S-Mustard gas **(Fig. 10)**. This ammunition was dumped in the Baltic Sea around Bornholm and south of Gotland, areas then under the control of the former Soviet Union. In the other occupation zones (USA, UK, France) a part from the total amount few details are available **(Fig. 7)**.

---

[13]Bundesamt (1993) and HELCOM CHEMU (1994). Arsenic oil is a mixture of diphenylchloroarsine and diphenyldichloroarsine.

[14]Tørnes (1997), Korzeniewski (1999) and Stock (1996) p. 53.

[15]Malyshev (1996), p. 94.

[16]HELCOM CHEMU (1993) 2/2/1 and HELCOM (1994).

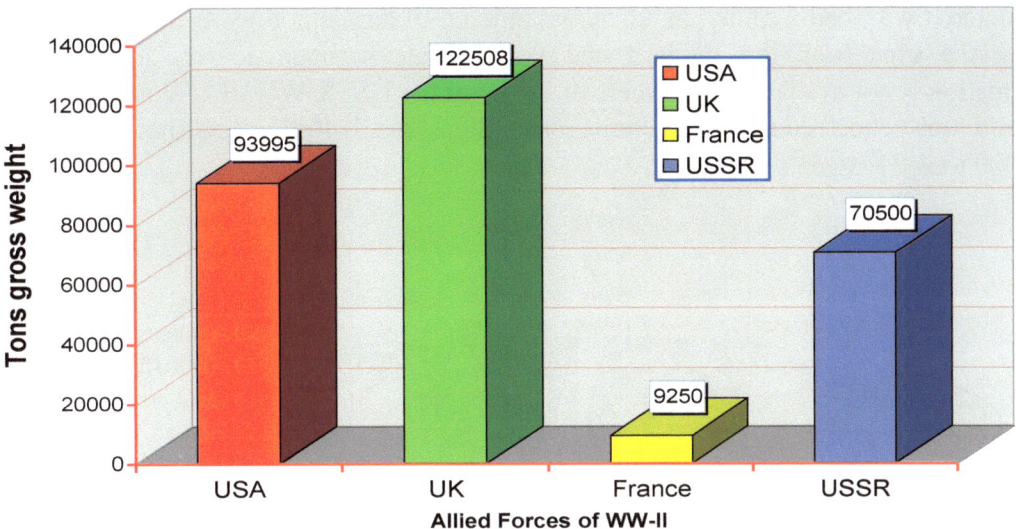

Fig. 7. Tons (gross weight) of CW ammunition discovered by the Allied Forces on German territory after WW-II.

3.3. Dumping strategies and location of dumping sites
3.3.1. Skagerak and North Sea

Dumping in the Skagerak was mainly carried out by the USA and UK authorities. 26 vessels[17] loaded with CW agents were sunk at a depth of 700 m 25 miles south-east of Arendal in the Skagerak **(Fig. 1)** with an estimated 130,000 tons of gross weight. Furthermore, 8 vessels were sunk to a depth of 200 m at a position west of Måseskår with a cargo of 20,000 tons of CW agents, the locations were: (a) $58^{\circ}05'$ - $58^{\circ}25'N$; $09^{\circ}15'$ - $09^{\circ}40'E$; (b) $58^{\circ}05'$ - $58^{\circ}15'N$; $10^{\circ}30'$ - $10^{\circ}55'E$.[18] Another 2 vessels (*Philip Heiniken and Marcy*) with 4500 tons CW agents were sunk in the North Sea (62° 57'N; 01° 32'E and 62° 59'N; 01° 23'E).[19]

3.3.2. Little Belt

Shortly before the end of WW-II, the Kriegsmarine[20] dumped 5000 tons of tabun and phosgene chemical ammunition and sunk 2 vessels loaded with 69,000 tabun grenades in the Little Belt at a depth of 30 m. In 1959/1960, the German authorities retrieved these 69,000 grenades. They were cast in concrete and sunk west of the Bay of Biscay.

3.3.3. Baltic Sea

Dumping in the Bornholm and Gotland basins was carried out by the Soviet authorities (**Fig. 1)** in the squares: $55^{\circ}07'$ - $55^{\circ}23'N$; $15^{\circ}28'$ - $15^{\circ}55'E$; $55^{\circ}56'$ - $56^{\circ}16'N$; $18^{\circ}39'$ - $19^{\circ}15'$ E, respectively. Only 4 vessels were sunk with their cargo of 15,000 tons of CW agents.[21] The rest was ammunition of which 35,000 tons were discarded overboard east of Bornholm at a

---

[17]HELCOM CHEMU (1994), p. 14-15.

[18]HELCOM (1993) 2/2/5.

[19]Frondorf (1996).

[20]Theobald and Rühl (1993).

[21]Theobald and Rühl (1993).

depth of 90 m while 2000 tons were dumped south of Gotland at a depth of 100 to 130 m.[22] The dumping was scheduled to be complete by the end of 1947. Due to difficulties encountered with bad weather, an unknown amount of ammunition was dropped overboard during the trips from the Wolgast and Peenemünde harbours to the dumping sites.[23] Information is not available as to where one third of the USSR WW-II CW ammunition was dumped, this concerned about 8500 tons mustard gas and 2240 tons lewisite of all produced 122,400 tons CW agents (**Fig. 11**).[24]

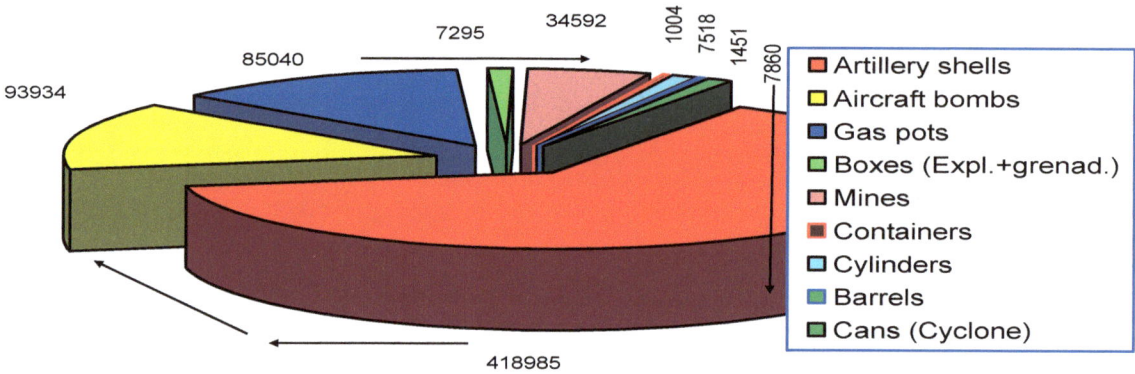

Fig. 8. Amount (in numbers) of CW ammunition dumped in areas of the Baltic Sea controlled by the former USSR.[25]

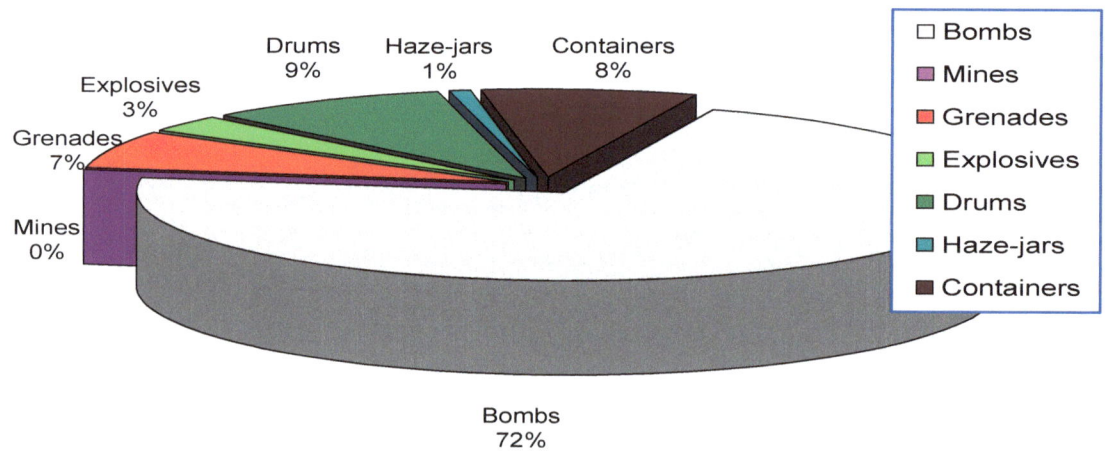

Fig. 9. Amounts (in %) of CW ammunition dumped in the Bornholm Basin.

[22]Bundesamt (1993).

[23]Bundesamt (1993), p. 11.

[24]TRUD (1999) Newspaper article, January, 1999 and Federov (1994).

[25]Official Russian document on the quantity of dumped chemical weapons in the Baltic and the Far East coastal waters after WW-II. Provided by Major General B.T.Surikov (in Russian), pp. 7.

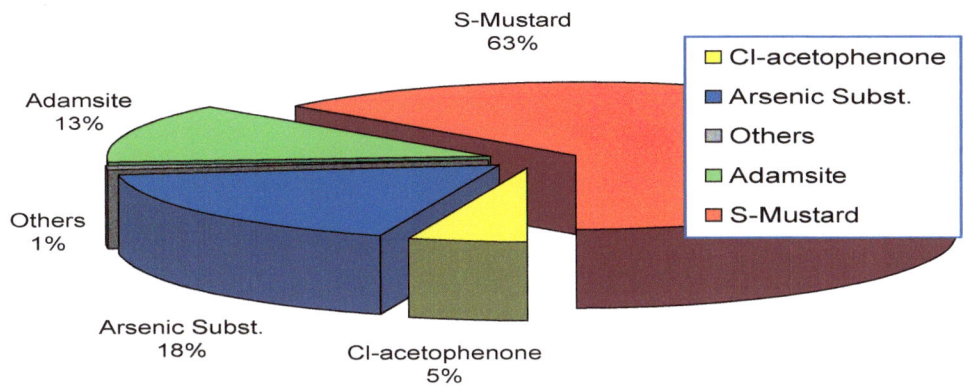

Fig. 10. Amounts (in %) of CW agents dumped in the Bornholm Basin. Cl-acetophenone is a tear gas and not really considered as a CW agent.

3.4. Restricted anchoring and fishing areas on nautical charts

Although the HELCOM CHEMU reports mention restricted areas (**Fig. 1**), these areas are not specified on the recent nautical charts (Annex IV) from the International Hydrographical Bureau of Monaco. However, areas indicated as being forbidden for anchoring and fishing, are located in the Little Belt and east of Bornholm where the chart alludes to CW agents (*Krigsgas*). The Gotland dumping site is characterised as *Ammunitionstipplats*. For the Skagerak, not all wrecks loaded with CW agents have been localised in the area as given by the nautical charts. Their positions are given in Table IV of Annex III and Chart D&E of Annex IV.[26]

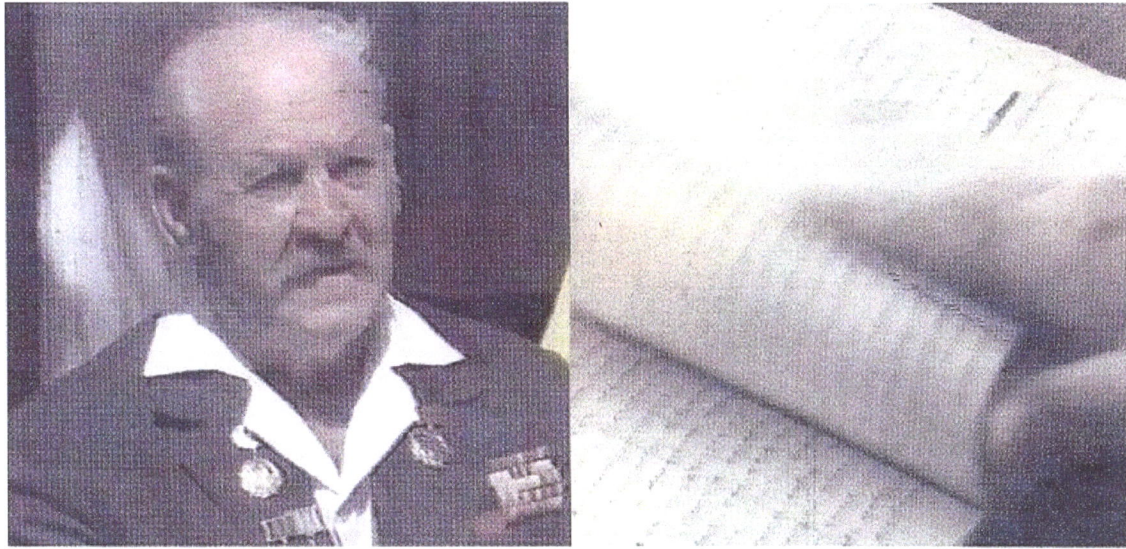

Fig. 11. Left. Captain Constatin P. Treskov, Lieutenant Commander of Soviet dumping in the Baltic Sea in August 1947. **Right.** His log-book of dumping. Photographs reproduced with permission from the Video flim "Cains's Smoke", Creative Ass. Film Program - the XXth Century Moscow.

---

[26]Tørnes (1997).

3.5. Corrosion

The CW bombs had a diameter between 20 and 48 cm[27] and a length of 109 - 181 cm. Their wall thickness ranged between only 1.5 and 3 mm (6 mm according to Surikov, 1999).

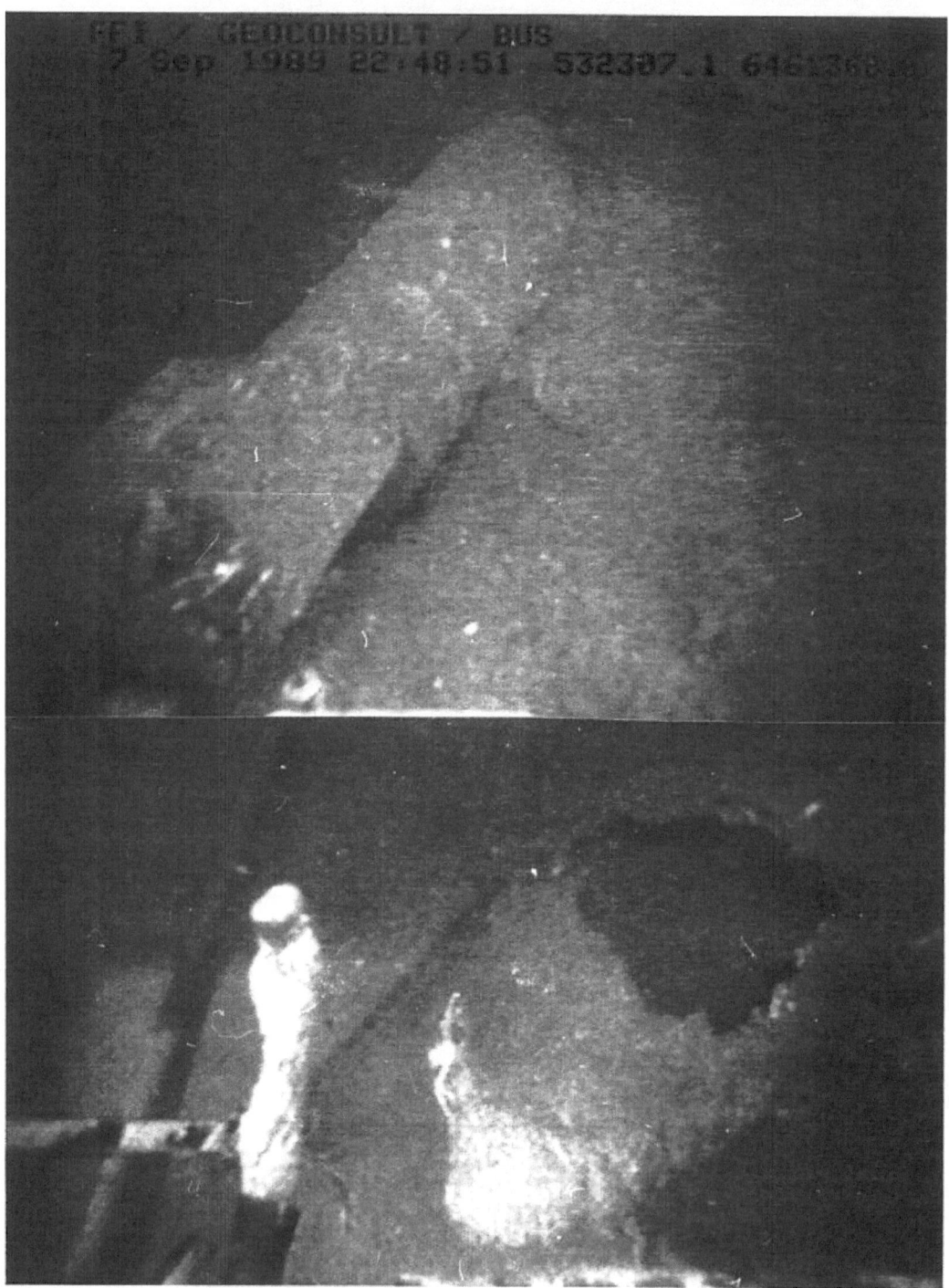

Fig. 12. State of corrosion of CW aircraft bombs located near a wrecked ship in the Skagerak. Photographs obtained by the courtesy of the Norwegian Defense Research Establishment (Dr. John Aa Tørnes).

---

[27]Bundesamt (1993), p. 15.

The 75, 105 and 150 mm artillery shells had thicker walls, of 8.9, 10.6 and 13.1 mm, respectively. CW drums, canisters and containers fall into the same category as the bombs, having a 2-3 mm wall thickness for the 10-150 kg containers and 5-6 mm for the heavier containers.[28]

According to Lisichkin,[29] citing the work of Kuntsevich and Evstafyev, 100% of the containers (barrels, drums) and certain aircraft bombs would have already undergone sealing loss as a result of corrosion. Considering the conditions on the seabed, it is likely that holes would progress by 0.15 mm a year.

Inspection by the Munitionsraumgruppe Schleswig-Holstein and the Bundeswehr[30] in 1971 showed that the investigated 28 bombs and 15 grenades in the Little Belt were covered by 50 cm slick and were heavily corroded. They did not contain either tabun or phosgene. It was also impossible to trace leaked CW agents in the neighbouring sediment. On the other hand, an investigation by the Norwegian Defense Research Institute[31] in 1989 at the dump site in the Skagerak, showed that both intact and corroded bombs were found and that some had corrosion holes (**Fig. 12**).

3.6. Properties of CW Agents

Danger for the European population of dumped CW agents arises when containers, drums and grenades leak due to corrosion. Potentially, the dumped 27,000 tons (net weight) of mustard gas had a capacity to kill 5.4 billion people, considering that the lethal dose per man is five gram[26] or 1.5 gram in one m^3 for one minute inhaled (see **Table IA**). If one thousandth is left, the killing capacity is still 5.4 million people.

Dissolution and auto-destruction

The chemistry of the CW agents is well described in chemical handbooks (See for further information Annex I). The environmental effects depend on two major factors:
* (i) their solubility in sea water and
* (ii) their rate of hydrolysis (decomposition) in sea water. This is expressed by the term half-life ($T_{1/2}$ = time in which half the amount has disappeared).

In **Fig. 13** some data are presented. However, mustard gas can also escape in leathery lumps, the inner parts of which remain reactive and do not lose their toxicity for several years.[32]

The major question concerning the fate of CW agents in the sea is: How fast will a CW agent dissolve in sea water and dissipate as harmless secondary products? Although on a laboratory scale this question can be answered, the real problem is how to relate these results to a marine environment,[33] where released CW agents can be transported as liquids **(Fig. 14)**, or be adsorbed by sediments and transported with moving silt and clay, or stick to fishing nets and thus reach fishermen or accumulate in marine organisms due to their affinity to dissolve in lipids.

---

[28]Surikov (1999) Russian information.

[29]Lisichkin (1996), p. 123.

[30]Bundesamt (1993), p. 15.

[31]Tørnes, Blanch, Wedervang, Andersen and Opstad (1989).

[32]IMO 1991.

[33]Federov (1995), p. 109-115.

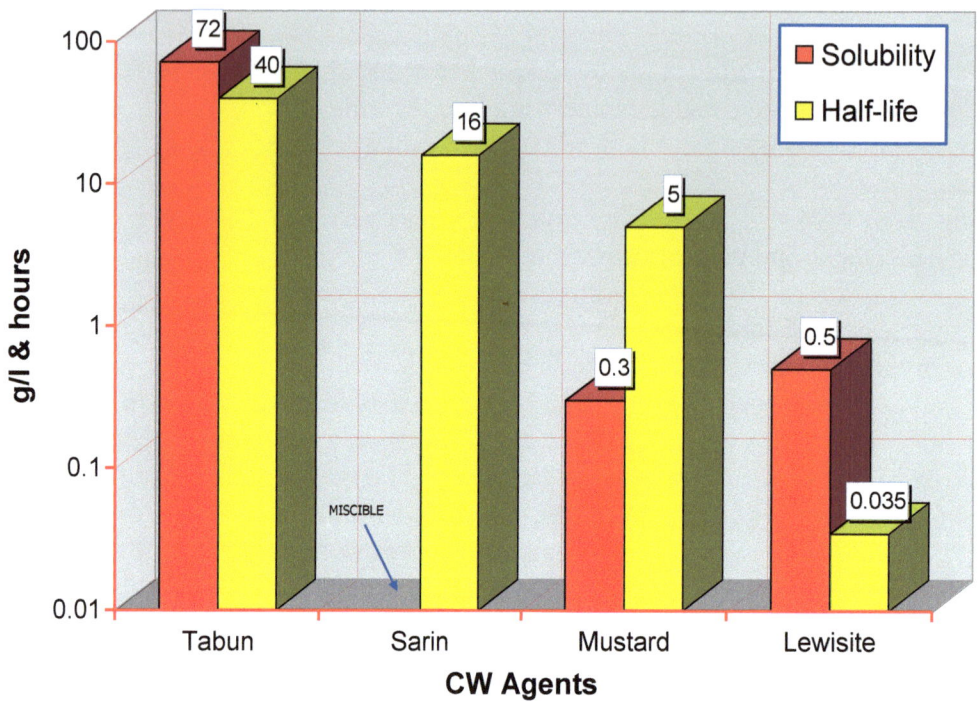

Fig. 13. Solubility and half-life in sea water of four major CW agents.

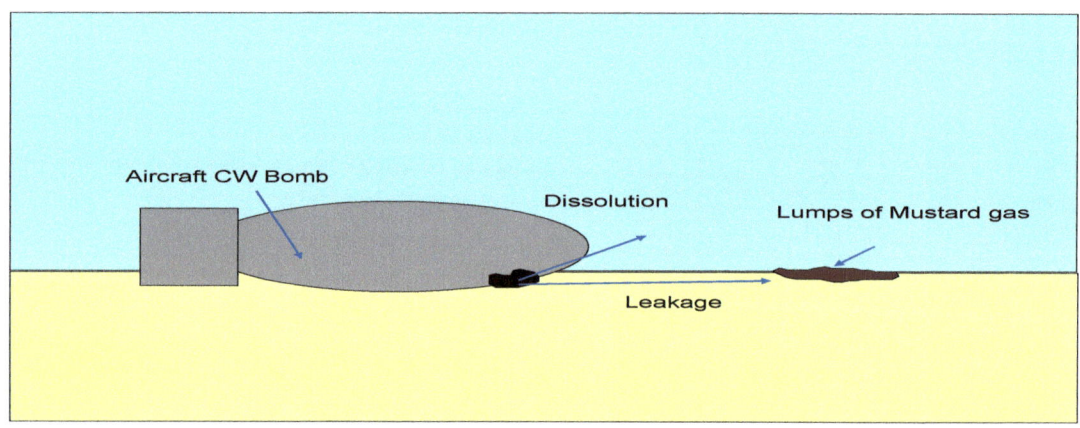

Fig. 14. Schematic illustration of an Ocean Dump Site and release of liquid CW agents with subsequent slow dissolution.

In spite of the fact that several model scenarios have been developed to predict environmental effects of the spreading of released CW agents,[34] few measurements have so far been published relating to non-arsenic CW gases in sea water around the dump sites. From some areas, close to the sunken vessel *Seostris* in the Skagerak, the water samples, analysed for mustard gas, tabun and thiodiglycol, showed negative results. This means that these CW agents were absent or occurred below the detection limits of respectively 200 ng/l, 2 μg/l and 10 μg/l (1 ng = 1 part per billion of a gram; 1 μg is 1 part per million of a gram).[35]

---

[34]MEDEA (1997).

[35]Tørnes (1992).

For arsenic, the concentrations determined in the western Baltic Sea in 1987[36] (0.2 to 0.8 µg/l) were close to natural concentrations in sea water. In comparison, the arsenic concentrations in the English Channel and the East Atlantic are between 2 and 3 µg/l.[37] This means that so far only small amounts of arsenic have leaked from the dumped CW material. A simple calculation (Annex II) demonstrates that if the total amount of 4322 tons arsenic (as a dissolved element) was to be liberated, the concentrations of arsenic, if distributed equally in one instant over the Baltic would increase by 0.2 µg/l. This may be very different above and around the dump sites, considering the weak currents and the slow renewal of the Baltic sea water by inflow of freshwater and exchange through the Kattegat and Little Belt. Thus there is every reason to ensure a slow leakage of arsenic CW agents into the sea.

4. LEGAL RESPONSIBILITY OF STATES
J.C.Duursma

4.0. International legal view
The OPCW may offer its good services by providing a venue for voluntary cooperation among the governments, relevant industries, academia and the NGO community, on the issues related to sea-dumped chemical weapons; (RC-3/NAT.14; 08-08-2013). The OPCW then acts as a venue for voluntary cooperation on sea-dumped chemical weapons.
For practical purposes and sea fishing industry in the Baltic Sea, the Commission of the EC stated in 2009: "In the Baltic Sea Region, hazardous substances continue to be a risk for the environment and for health. They include organic contaminants and heavy metals, as well as chemical weapons sunk in the Baltic Sea. Once released into the sea, hazardous substances can remain in the marine environment for very long periods and accumulate in the marine food web." (SEC (2009) 712/2, p.15)
Taking into account earlier work by HELCOM, the EC tries to protect the environment and fishing industry in the Baltic Sea and assesses the need to clean up chemical weapons.
On 20 December 2010, the UN adopted a Landmark Text on Protecting Coral Reefs, Mitigating Ill Effects of Chemical Munitions dumped at Sea Res. It encourages cooperative measures to assess and increase awareness of environmental effects related to waste originating from chemical munitions dumped at sea; (GA/11040)/
The scientific community and civil society, including the International Scientific Advisory Board on Sea-Dumped Chemical Weapons and the International Dialogue on Underwater Munitions, have achieved this with a very good cooperation among all United Nations Member States and remarkable input.

4.1. Introduction
Chemical munitions which have been dumped at sea by order of the Allied States after the end of WW-II may sooner or later damage the marine environment. The legal question then arises of which State is internationally responsible for the damage resulting from the dumping and which State has to pay compensation to the injured claimant. According to public international law, a State can only be held internationally responsible for an act which, at the time it was performed, constituted a breach of an international obligation which

---

[36]Theobald and Rühl (1993), p. 128.
[37]Riley and Skirrow, 1965, p. 325.

was in force for that State at that time.[38] One should therefore first determine what constitutes a possible illegal act, second which international legal obligation has been violated and third which State is internationally responsible for the breach.

In the present case, one may distinguish three acts which could lead to the international legal responsibility of the State to whom the acts can be attributed, namely:

1. the dumping of the chemical munitions at sea at the end or after WW-II,
2. the failure to destroy the dumped chemical munitions,
3. the failure to protect and preserve the marine environment by not taking the necessary steps to prevent harm from the dumped chemical munitions.

As will be demonstrated below, it is most likely that, in the present case, only the third act (failure to protect and preserve the marine environment) may entail the international legal responsibility of the States in whose territorial waters or Exclusive Economic Zone (EEZ) the chemical weapons have been dumped. The first and second acts (dumping and failure to destroy chemical weapons) are closely linked and it is unlikely that any State will be held internationally responsible for them.

4.2. Responsibility for dumping and destruction

At the end of WW-II and immediately thereafter, the dumping of chemical waste in the high seas (i.e. beyond the territorial seas) was not prohibited by international law, but considered a "freedom of the high seas". At the time, the law of the sea was based on international customary law, which only later, in 1958, was codified in multilateral treaties.[39] The dumping in the high seas was legal if reasonable regard was paid to the interests of other States in their exercise of the freedom of the high seas.[40] After WW-II the disposal of German chemical weapons by dumping at sea was considered a just and viable solution to prevent any short-term harm to human health.[41] Even now, the States Parties to the 1974 Convention on the Protection of the Marine Environment of the Baltic Sea Area (hereafter, the Helsinki Convention) do not prohibit dumping in the Baltic Sea Area "if dumping appears to be the only way of averting the threat and if there is every probability that the damage consequent upon such dumping will be less than would otherwise occur."[42][43] The dumping of the German chemical munitions in the territorial seas by the Allies would have necessitated approval by the coastal States to avoid being considered an act of aggression.[44] There does not seem to be any proof that the coastal States objected to the dumping after WW-II or that they maintained any protest against the Allied States for the dumping of German chemical munitions in their territorial seas.

---

[38] Art. 18(1) Draft Articles on State Responsibility, 37 International Legal Materials (hereafter I.L.M.) (1998), p. 446.

[39] The Convention on the Territorial Sea and the Contiguous Zone 1958, the Convention on the High Seas 1958 and the Convention on the Continental Shelf 1958.

[40] Principle codified in Art. 2 Convention on the High Seas; Timagenis (1980), p. 114.

[41] The dumping carried out by the Allies met with the regulations of the Potsdam Conference (2 August 1945): Korzeniewski (1994), p. 91.

[42] The Baltic Sea Area comprises the Baltic Sea proper with the Gulf of Bothnia, the Gulf of Finland and the entrance of the Baltic Sea bounded by the parallel of the Skaw in the Skagerak at 57o44'8"N: Art. 1 Helsinki Convention 1974; 13 I.L.M. (1974), p. 546.

[43] Art. 9(4) Helsinki Convention 1974, ratified by Denmark, Finland, Germany, Poland, Russia and Sweden.

[44] At the time, the breadth of the territorial sea was established by each State separately: Denmark and Norway 4 nautical miles (1745) and Sweden also 4 nautical miles (1779): Brownlie (1979), p. 191, n. 5. At present the breadth of the territorial sea should not exceed 12 nautical miles: Art. 3 of UNCLOS III (1982).

It has been reported that some Baltic States were pressed to accept the dumping in their territorial seas by the Allied Forces or were not informed of the dumping operations.[45] However, legally speaking the consent of the Baltic States is established as the coercion by the Allied Forces was not illegal - they did not threaten to use force against the Baltic States to make them approve the dumping.

The legal argument of necessity to dispose of the chemical weapons as safely as possible in, at that time, the most practical way indicates that the interests of the coastal States involved had been taken into account. As the dumping of the German chemical munitions at sea was considered legal at the time of the dumping, this cannot lead to the international responsibility of the dumping (Allied) States under present international law.

The international legal obligation to destroy these dumped chemical weapons however is another matter. Since the entry into force on 29 April 1997 of the Convention on the Prohibition of the Development, Production, Stockpiling and Use of Chemical Weapons and on their Destruction of 1993 (hereafter, the OPCW Convention) the States Parties are in principle obliged to destroy old chemical weapons dumped at sea that have been confirmed by the OPCW Technical Secretariat as meeting the definition of ☐old chemical weapons☐ not later than 10 years after the Convention has entered into force for them.[46] Old chemical weapons are defined as: chemical weapons which were produced before 1925 or chemical weapons produced in the period between 1925 and 1946 that have deteriorated to such extent that they can no longer be used as chemical weapons,[47] therefore including the German dumped chemical munitions under discussion.

This duty to destroy old chemical weapons is limited to the old chemical weapons that are located on the territory of the State Party which includes the territorial sea, but excludes the Exclusive Economic Zone (hereafter, EEZ). Only these old chemical weapons have to be declared and will be the object of inspections by the Technical Secretariat, leading to confirmation of conformity with the definition of old chemical weapons and eventually destruction.

A State has to declare to the Technical Secretariat of the OPCW whether it has on its territory old chemical weapons and provide all available information.[48] However, the duty to declare old chemical weapons and to destroy them can be waived unilaterally at the discretion of a State Party for chemical weapons which have been dumped at sea before 1 January 1985.[49] The drafters of the OPCW Convention have thus voluntarily chosen to give States Parties the choice between destroying (the sea-dumped German chemical weapons among others) and monitoring the *status quo*.

Well aware of the technical and financial difficulties to destroy such old chemical weapons and considering that such a destruction would not be of primary security interest to the States Parties, the coastal States have been left the freedom not to declare and destroy the

---

[45]Knightley (1992), writing about the post WW-II dumping operations in the Baltic Sea ☐(...) the Soviet occupation force chose an area east of the Danish island Bornholm (...). The Danish Government was not told of the operation, which was concluded in the last week of December 1947 (...) according to a Norwegian department of environment report of 1986, Norway was forced to accept the dumping under Anglo American pressure. Sweden and Denmark, although they have not said officially, were also pressed to accept the dumps.☐

[46]Part IV(B)(7) OPCW Convention, 32 I.L.M. (1993) p. 804. Denmark, Germany, Norway, Russia and Sweden are parties to the Convention.

[47]Art. II(5) OPCW Convention .

[48]Art. III(1)(b)(i) OPCW Convention .

[49]Arts. III(2) and IV(17) OPCW Convention. The cut-off date of 1 January 1985 is probably related to the dumping activities of one of the major possessors of chemical weapons, the USA, which reportedly dumped nerve gas, explosives and radioactive waste in the Atlantic Ocean in the sixties and seventies. See Krutzsch and Trapp (1994), p. 58, Soni (1985), p. 215 and Timagenis (1980), p. 115.

sea-dumped chemical weapons of WW-II. The environmental threats of such dump-sites however are not considered by the OPCW Convention. Yet, if these old chemical weapons are recovered and brought to land they may no longer be considered "dumped at sea before 1 January 1985", but fall under the general obligation to destroy old chemical weapons.[50] It is thus in the interest of States to leave the old chemical weapons at sea. Another legal question concerns the declaration and destruction of abandoned chemical weapons, meaning "chemical weapons, including old chemical weapons abandoned by a State after 1 January 1925 on the territory of another State without the consent of the latter".[51] In the event that the coastal States did not at the time of dumping after WW-II implicitly or explicitly approve the sea-dumping, then the old chemical weapons will fall under the definition of "abandoned chemical weapons", the destruction of which should be undertaken by the State Party who dumped the weapons, i.e. the Allied forces.[52] The freedom not to declare and destroy chemical weapons dumped at sea before 1 January 1985 does not legally apply to these abandoned chemical weapons.[53] One notices that the fact that the old or abandoned chemical weapons used to be owned or possessed by Germany (or any other State) is irrelevant to the international legal obligation under the OPCW Convention to declare and destroy these weapons.[54] As the coastal States involved are presumed to have consented to the dumping at sea of the WW-II chemical munitions, no State can be legally obliged to declare and destroy these munitions.

4.3. Responsibility for the protection and preservation of the marine environment

On 22 March 1974, the 7 States surrounding the Baltic Sea[55] signed the Helsinki Convention which obliges the States Parties, among others, to "take all appropriate legislative, administrative or other relevant measures in order to prevent and abate pollution and to protect and enhance the marine environment of the Baltic Sea Area".[56] This obligation implies that the coastal States have to take all appropriate measures to protect the marine environment (including human health)[57] from pollution caused by the leakage from the dumped chemical munitions in the Baltic Sea Area (which stretches beyond the territorial seas of the coastal States). The circumstances of each case will determine whether the measures taken are "appropriate".[58] The Helsinki Commission (established for the purposes of the Helsinki Convention) therefore created the *ad hoc* Working Group on Dumped Chemical Munitions (HELCOM CHEMU) in 1993. The reports and recommendations of

---

[50]Stock (1996), p. 65.

[51]Art. II(6) OPCW Convention .

[52]Art. I(3) and Part IV(B)(8-18) OPCW Convention.

[53]Krutzsch and Trapp (1994), p. 58 and 76.

54Art. IV(1) OPCW Convention explicitly excludes old chemical weapons and abandoned chemical weapons from the provisions of Art.IV. Therefore Art. IV(11) does not apply to these weapons, although it states: ▯Any State Party which has on its territory chemical weapons that are owned or possessed by another State, or that are located in any place under the jurisdiction or control of another State, shall make the fullest efforts to ensure that these chemical weapons are removed from its territory not later than one year after this Convention enters into force for it.▯

[55]Denmark, Finland, GDR, FRG, Poland, Sweden and the Soviet Union.

[56]Art. 3(1) Helsinki Convention.

[57]"Pollution" as defined by Art. 2(1) of the Helsinki Convention means: "introduction by man, directly or indirectly, of substances or energy into the marine environment, including estuaries, resulting in such deleterious effects as hazard to human health, harm to living resources and marine life, hindrance to legitimate uses of the sea, including fishery, impairment of the quality for use of sea water, and reduction of amenities."

[58]If there is a dispute between the parties as to the interpretation of this term, they have to settle the matter by negotiation, good offices or mediation of a third contracting party. By agreement they may decide to submit the case to an arbitral tribunal or the International Court of Justice, Art. 18 Helsinki Convention.

HELCOM CHEMU are not legally binding on the contracting States, but may serve as a guideline for what can be considered "appropriate measures". Thus, the protection and preservation of the Baltic Sea Area from pollution by dumped chemical munitions will be appropriately secured, according to the HELCOM CHEMU reports if:

1. Chemical munitions caught by fishermen and presenting a possible risk to human health are redeposited in the sea with great caution and following the instructions of the appropriate authorities in co-ordination with the principles and procedures, agreed upon at the Ninth Meeting of the Helsinki Commission.[59]
2. The chemical munitions from the Baltic Sea Area are not recovered.[60]
3. The contracting States follow the HELCOM Guidelines to Authorities in dealing with Chemical Munitions caught by Fishermen.[61]
4. The contracting States provide Denmark with information on all national and international activities concerning dumped chemical munitions by the end of June every year (including reports on catches of chemical munitions by fishermen and scientific and practical studies of their effects on the marine environment).[62]
5. The contracting States carry out further investigations for the location and characterisation of the dumped chemical munitions as well as their ecological effects.[63]

In case the dumped chemical munitions cause harm to human health it is unlikely that a State Party to the Helsinki Convention can be held legally responsible if it has implemented the above recommendations of the HELCOM CHEMU as it will have taken all appropriate measures according to the Helsinki Convention.

This conclusion stands apart from the responsibility of States by virtue of UNCLOS III.[64] Under this Convention: "States shall take individually or jointly as appropriate, all measures consistent with this Convention that are *necessary* to prevent, reduce and control pollution of the marine environment from any source, using for this purpose *the best practicable means at their disposal and in accordance with their capabilities* and they shall endeavour to harmonise their policies in this connection [emphasis added]".[65]

One notices that in UNCLOS III too, as in the Helsinki Convention, no absolute duty to achieve a fixed end result is imposed, but the duty is limited within a framework of reasonableness. In the present case this means that the prevention, reduction and control of pollution from the dumped chemical weapons at sea can only be legally demanded if it is necessary, if the best practical means of which the State can dispose are used and if it is within the technical and financial capabilities of that State. In any case, the States Parties will have to co-operate on a global or regional basis in order to formulate and elaborate international rules, standards and recommended practices and procedures to protect and preserve the marine environment.[66] The duty to co-operate is absolute and the Helsinki Convention and recommendations of the HELCOM CHEMU are a reflection of this. Furthermore, the duty to co-operate includes a duty for States Parties to jointly develop and

---

[59]HELCOM CHEMU (1994) Conclusion No. 4., p. 42.

[60]HELCOM CHEMU (1994) Conclusion No. 12., p. 43.

[61]HELCOM CHEMU (1995), para 3.5, p. 60.

[62]HELCOM CHEMU (1995), p. 5. In his letter of 25 January 1999, Mr. Kjell Grip, Secretary of the Protection Commission of the Helsinki Commission stated: As lead country Denmark has to give a yearly report to the Environment Committee. For 1998 there was nothing to be reported. (Remark editor): The incident with the vessel Prof. Shtockman has not yet (by June 1999) been reported to Denmark, see above at 1. Introduction.

[63]HELCOM CHEMU (1995), pp. 5 and 6.

[64]The Helsinki Convention is without prejudice to UNCLOS III: Art. 21 Helsinki Convention.

[65]Art 194(1) UNCLOS III. Germany, Norway, Russia and Sweden are parties to UNCLOS III, Denmark is not.

[66]Art. 197 UNCLOS III.

promote contingency plans for responding to pollution incidents or in case of imminent danger of damage to the marine environment.[67] As it is not entirely excluded that the dumped chemical munitions will leak and constitute a potential threat to the marine environment, the States Parties to UNCLOS III are under the obligation to prepare contingency (emergency) plans. These can be used at the moment the emergency arises. In this case the duty to draft emergency measures is not limited by arguments of "reasonableness", in contrast to the duty to co-operate in eliminating the effects of pollution and preventing or minimising the damage, which is less stringently formulated.[68] Contingency plans can take the form of international agreements. The failure to jointly develop and promote such plans for risk areas such as dump sites entails a State's international legal responsibility under UNCLOS III.

A contingency plan for emergency situations caused by dumped chemical munitions could be included in a proposed Protocol to the Helsinki Convention, specifically dealing with the protection of the marine environment against CW agents.[69]

UNCLOS III also obliges States Parties to co-operate in the implementation and further development of international law relating to responsibility and liability for the assessment of and compensation for all damage caused by pollution of the marine environment.[70] This includes, where appropriate, the development of criteria and procedures for payment of adequate compensation, such as compulsory insurance or compensation funds.[71] The duty to co-operate in order, among other things, to develop international rules relating to the compensation for damage caused by dumped WW-II chemical weapons is not a hard one. Inherently a duty to co-operate is no obligation to achieve a result, only *to try* through joint action by States to come to a result. Only if the States Parties to UNCLOS III do not try to further develop international law relating to liability for all damage caused by pollution of the marine environment can they be held internationally responsible by other State Parties.

States who are not parties to UNCLOS III, such as Denmark, will only be bound by the provisions of this Convention which are considered international customary rules such as the duty to prevent, reduce and control pollution of the marine environment.[72] The international customary legal obligation to prevent environmental harm to other States includes the procedural obligation to inform and to consult and negotiate regarding activities presenting a risk of environmental injury.[73] In other words, even if a State is not a party to UNCLOS III and chemical munitions dumped in its territorial sea or EEZ start leaking and present a threat to the marine environment, that State has to prevent pollution and inform other States at risk. If it does not take the appropriate measures, other States can invoke its international legal responsibility and claim compensation for the damage caused.

4.4. Conclusions

Under present international law only the State in whose territorial sea or EEZ the WW-II chemical weapons have been dumped can be held internationally liable for damage caused to the marine and human environment by the leakage of toxic munitions if that State has not taken the appropriate measures to prevent the pollution. The term ▯appropriate▯ is not static.

---

[67] Arts. 198 and 199 UNCLOS III.

[68] Art. 199 UNCLOS III provides "(...) States in the area effected, <u>in accordance with their capabilities</u> (...) shall co-operate, <u>to the extent possible</u>, in eliminating the effects of pollution and preventing or minimising the damage. To this end, States <u>shall</u> jointly develop and promote contingency plans for responding to pollution incidents in the marine environment [emphasis added]."

[69] See: Conclusions, Warnings and Steps to be Taken.

[70] Art. 235(3) UNCLOS III.

[71] *Ibid.*

[72] Art. 194 UNCLOS III. See Smith (1988), p. 75.

[73] Smith (1988), p. 81.

The more is known about the environmental effects of the dumped chemical weapons and the more technical knowledge is available to a coastal State to prevent marine pollution from the dumped munitions, the more stringent is the duty to actively protect and preserve the marine environment. What is appropriate now, may no longer be considered appropriate in the future. A sustained co-operation between States to tackle environmental threats from dumped chemical weapons is legally imposed. The measures to be taken must be as far reaching as possible.

This obligation of coastal States is the only international legal safety net for the presentation of the marine and human environment as, in the present case, neither Germany (former owner of the chemical weapons) nor the Allied Forces (the dumping States) can be held liable. Moreover, the OPCW Convention leaves the coastal States every freedom not to declare and destroy the old sea-dumped chemical weapons in question. Thus, the environmental effects of the dumped WW-II chemical weapons have to remain the object of international legal attention in the future.

5. TECHNICAL OPTIONS

Whereas a well-equipped vessel with modern sub-marine devices could effectively treat 10 munitions per day, it would take 165 years to retrieve the 600 thousand grenades, bombs and other CW ammunition (67,400 tons[74] gross weight, based on Russian reports) spread over 250 km^2 in the Baltic Sea. Considering the additional amount of the 168,000 tons[75] dumped in the Skagerak, this is obviously an impossible task.

Should we therefore abandon any active measures, in spite of possible risks in the near future for sailors and coastal populations, and leave the Coastguard, Navy and Coastal Rescue authorities to improvise in emergency situations?

The greatest risks are:
* the leakage of lumps of S-mustard from dumped ammunition in the Baltic Sea
* and the contamination by CW agents from collapsed corroded cargo in the wrecked ships of the Skagerak.

A number of options are presented based on the confinement of wrecked ships and burial of ammunition at risk and the destruction of lumps of mustard gas captured and retrieved by fishing, research or exploration vessels at the different sites.

5.1. Confinement of wrecked ships and CW cargo
J.G.de Vries
The danger of leakage and spreading of CW agents from the wrecked ships loaded with CW ammunition should not be underestimated. Due to corrosion of the aircraft bombs, this ammunition will collapse under their weight and that of the load of grenades. Tidal currents and turbulence caused by the passage of trawl nets and submarines may cause additional damage. In spite of the warnings on nautical charts, it was found that some of the wrecked ships in the Skagerak were covered by trawl nets.[76]

---

[74]Stock (1996), p. 52.
[75]Stock (1996), p. 53.
[76]Granbom (1996), p. 42.

5.1.1. Remedial actions

A variety of techniques[77] can be suggested to retrieve CW ammunition such as:

Lifting of small objects

Many ROV's (Remotely Operated Vehicles) are active in the offshore industry and those containing specialised handling tools could be employed to place dumped ammunition in a disassembly container for dismantling and (pre-)processing with electro- or thermo-chemical methods on land or on a mother ship (see section 4.2.)

Lifting of ships

This is difficult for several reasons: the corroded state of the vessels and their cargo, their weakened structure caused by holing the ship before sinking, their great weight and the risk of explosions which would release toxic gas when surfacing (these ammunition explosions may occur due to pressure changes between the seabed and the surface).

5.1.2. Techniques of sarcophaging

Pre-survey

A method, based on existing techniques and equipment is proposed to sarcophage sunken vessels and dumping sites with a sheet of granular material. Since the material is normally crushed rock, supplied by quarries, it is hereafter simply denominated as rock.

Prior to the installation of a rock protection, a survey has to be made to determine the seabed and sub-bottom conditions:
*   soil sampling: by gravity cores, vibro-cores, cone penetration and grab samplers
*   buried objects: by sub-bottom profiler and magnetometer
*   seabed: by echo-sounder and side-scan sonar
*   visual inspection: by eye-ball ROV
*   location: by global and local positioning systems.

Data have to be collected on:
*   the surface to be covered, for instance in case of a ship, its length, width, height, position on, and, if partially buried, in the seabed and the condition of its structure.
*   the presence of objects on or buried in the seabed..

Advantages of sarcophaging

If properly applied sarcophaging will:
*   isolate the CW agents and prohibit dispersion over a wider area;
*   protect ammunition and shipwrecks from being damaged and ruptured by foreign objects such as bottom-towed fishing gear, anchors and exploration activities;
*   reduce the speed of the ammunition's corrosion as the water is almost stagnant and the influx of oxygen thus reduced;
*   retard the release of toxic agents when the material (or combination of materials) applied has a low fluid permeability;
*   allow dissolution and disappearance by hydrolysis of the CW agents to take place in a confined site;
*   allow chemical conversion of leaked CW agents when gels of reacting agents are added.

---

[77]There are a variety of Russian proposals for monitoring and confinement of CW ammunitions among which CONTEX (1995) and Kurganov and Morozov (1999).

5.1.3. Stone and sediment (rock) confinement
Techniques of sarcophaging

The method of full sarcophaging of a wrecked ship consists of building a stable mound of rock along and over the wreck. The quantity of rock needed is, for instance, 250,000 tons for a Liberty ship as used during WW-II (**Fig. 15**).

The amount of rock is significantly less when a wrecked ship is already buried in the seabed or when the wrecked ship is intentionally lowered into the seabed.

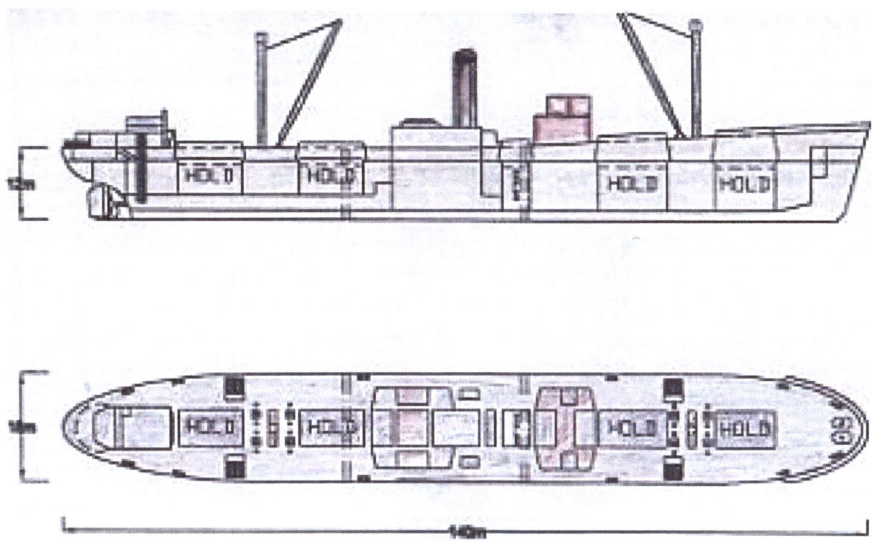

Fig. 15. Example of a WW-II Liberty ship wrecked with CW agents in the Skagerak.

Rock-covering

Whereas dumping of sand and rock was originally carried out by split-bottom and side-dumping vessels, a new generation of vessels was built in the late seventies, when offshore oil and gas installations entered deeper water and more difficult working conditions had to be overcome. These vessels are equipped with a fall-pipe system for controlled placement of the material using underwater cameras to ensure low losses during deposition (see **Fig. 16 , 17 and 18**).

These vessels navigate on a Dynamic Positioning System (DPS) and time-consuming anchor handling operations are not needed. The loading capacities of a recently built new generation of vessels range from 24,000 to 36,000 tons. The fall-pipes have a diameter between 0.5 and 1.0 metres. Some have an open structure (nets with circular rings at regular distances), others have closed-wall plastic or steel pipes, stacked on top of each other. The bottom end of the fall-pipe is located within the central opening of a cylindrical Remotely Operated Vehicle (ROV). With its thrusters **(Fig. 19 & 20),** the ROV holds the fall-pipe exit with precision above the required position.

The survey system of the ROV is linked into the vessel's DPS system, so that its position relative to the vessel and also its actual position are always known. The ROV is heave-compensated, which makes it independent of the ship's motion and it can be kept at a constant level above the seabed.

Fig. 16 – Rockdumping Vessel "Flintstone" of Tideway

Fig. 17 – Rockdumping Vessel "Willem de Vlamingh" of Jan de Nul

The vessels can operate over a large depth range. Projects have been carried out worldwide in such areas as the North Sea, offshore Mexico, The Philippines and Australia in depths up to nearly 1000 meters. The last generation of DP rock dumping vessels have been designed for accurate dumping up to 2000 meters.

In recent years these vessels have been applied with stone dumping in the Baltic for a very different purpose. The Nord Stream gas project **(Fig. 18)** consists of a major pipeline between Russia and Germany, which runs mainly over very irregular rocky area. To

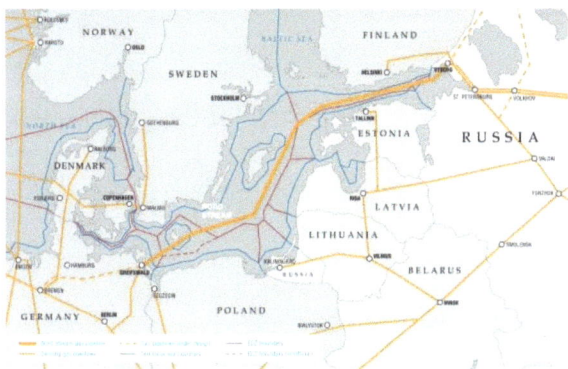

reduce the spans between successive rocky peaks (risk of vibrations), there were many additional supports of rocks deposited.

Fig. 18. Gas pipeline in the Baltic Sea. Figure available on Internet.

To ensure the integrity of the sand/rock-deposited structure on the seabed, at least two criteria have to be met:

* the stones in the outer layer, also called the armour layer, have to be stable under extreme weather conditions to provide the structure's permanent integrity;
* one or more intermediate layers, also called filter layers, are required when there is a big difference in particle size between the armour layer and the seabed sediments. Without such filters, there will always be a risk of seabed material being washed out through the pores of the outer layer and of the larger stones descending into the seabed.

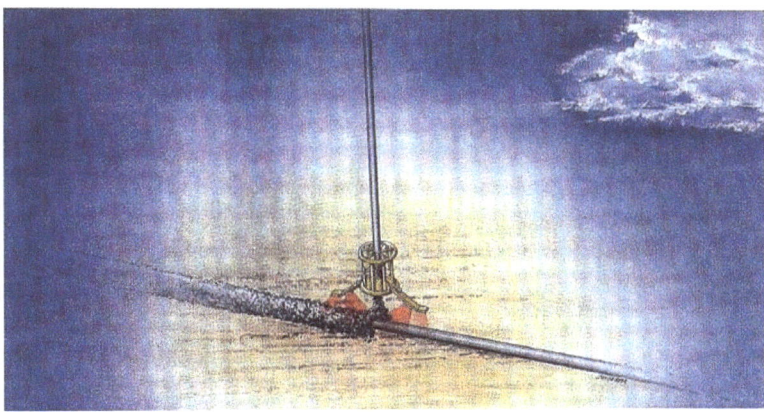

Fig. 19. The bottom end of the fall-pipe is connected to a Remotely Operated Vehicle (ROV).

Outer Armour Layer

The material of the armour layer should be sufficiently heavy to withstand severe conditions. For this purpose, the formulated design condition must allow for extreme weather conditions in terms of a maximum sea wave (amplitude, period and direction) and a maximum storm (speed and direction), with an annual occurrence probability of 0.01 times, also referred to as a return period of 100 years. Both have to be transformed into current velocities just above the seabed with wave-induced velocity being a continuously varying, oscillatory vector and the wind- or tide-generated current static vector.

When stones have to be deposited at sites 100 metres deep or more, the stone size can be kept rather small (< 75 mm), as only stability variations have to be taken into account. Where fishing activities are likely, it is better to use larger material with diameters of 75 to

125 mm. Since the material is the result of rock crushing, the given size of 75-125 mm generally ranges between 25 and 200 mm.

Intermediate Filter Layer(s)

It is an empirical rule for breakwaters, piers and groins that the grain-size ratio of the material in two successive layers should be less than 25. For offshore conditions, a less costly, in between solution can be applied, whereby the two layers are installed in a combined single operation. In this case the material for the filter and armour layers are mixed on land before loading. The disadvantage here is that some finer material is lost.

Permeability

Pressure gradients around a sarcophaged structure are the driving force of the ground water movements. The permeability of the used filter material can range from 1 m/sec for large stones to 0.4 mm/hour for sandy clay.[78] This lower value is acceptable for halting the release of CW agents.

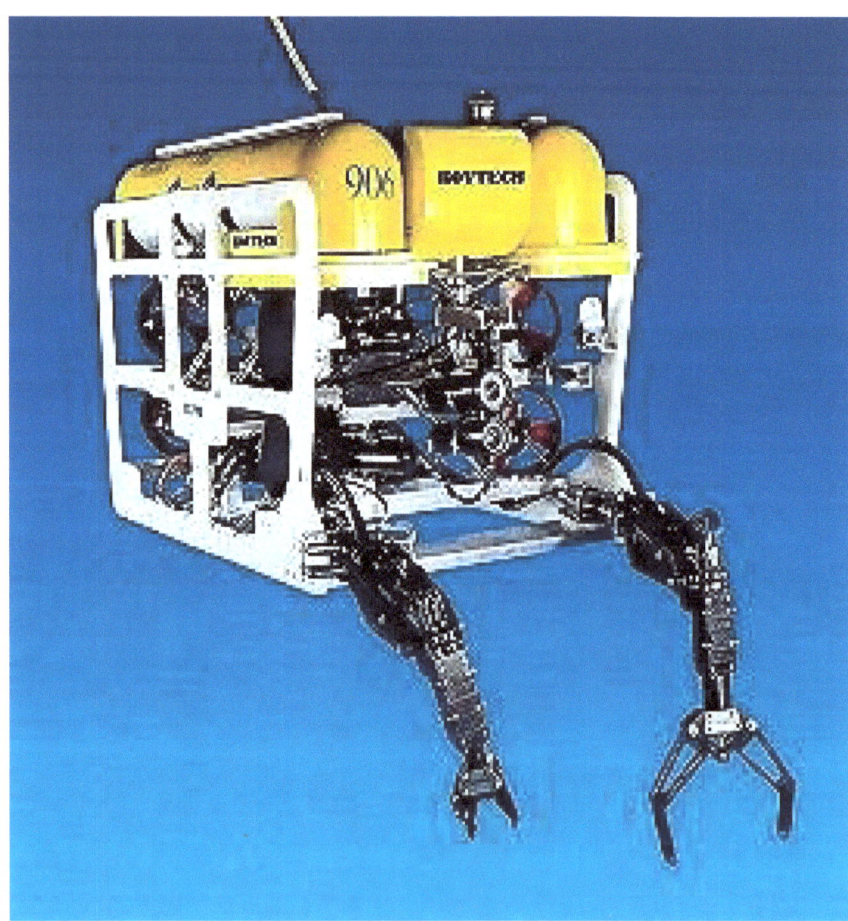

Fig.20. Remotely Operated Vehicle

There are various ways of reducing the permeability of the armour and filter protection layers:

* Installation of watertight flexible mattresses filled with grout cement or asphalt bitumen. Stones placed along the edges fill the scour holes and stop erosion.

---

[78]CIRIA (1991).

* Injection of grout cement, asphalt or polymer cement **after** sarcophaging. However these techniques have their technical complications: Grout requires precautions to avoid cracking and loss of pieces of concrete, and needs certain additives to prevent de-mixture during deposition. Asphalt has to be poured hot, which requires an insulated chute. This is practically impossible at depth, due to the cold sea-water environment at which the CW agents have been dumped. Polymer cement is applied as a thin liquid and requires an additive for quick hardening to avoid being flushed away. This means that there must be a well-designed application system to avoid interruption of the process. In high seas this is can be a problem!

* Application of geotextiles. These woven textiles are widely used for river banks and sea-defences. Geotextiles can be spooled on large drums and spread over a wide area in one single operation. Extra weight must be added to ensure efficient lowering onto the seabed and anchoring at the site.

* Application of additives which facilitate the hydrolysis of CW agents. As mentioned in Annex I on the chemistry of CW agents, hydrolysis is accelerated under alkaline conditions. Such conditions are created with additives such as chalk. The use of ground carbonate rock, if available, for the filter layer would perhaps be a solution.

* Injection of special gels. A gel called Khitozan,[79] produced from crab shales, mixed with fly ash from coal-fired power stations was tested to isolate radionuclides in the sunken Russian nuclear submarine *Komsomolets* on 7 April 1989 in the Norwegian Sea, 180 miles south-east of the island Medvezhiy. Certain technical problems have to be solved before applying fluids which have to solidify beneath the armour or filter layer. They may easily solidify in the supply hose. The gel material is also difficult to produce at a reasonable cost.

Fig. 21.A. *Hydrodigger,* **B.** Launching of the *Hydrodigger.* Illustrations reproduced by courtesy of Seafloor Dynamex, Howemoss Drive, Kirkhill Industrial Estate, Dyce Aberdeen AB21 OGL, U.K.

[79]Surikov (1996), p. 70.

Burial of wrecked ships

Recently a new generation of trailing suction hopper dredgers (TSHD) have been built which can dredge in depths of 150 m, their use in the Baltic Sea is therefore possible. Other, slower techniques are available for greater depths, such as in the Skagerak, using water jetting, hydro-dynamic excavation and deep-water dredging. In Fig. 18 a heavy working class ROV is given which can be equipped with various tools, such as a jet pump which produces a high pressure jet flow by which the sediment is removed. The system operates satisfactory to depths of 1700m.

Another applicable equipment (not an ROV) called *Hydrodigger* (**Fig. 21**), houses a large horizontal propeller which generates a massive down-ward flow of water, by which the seabed underneath is gradually eroded. An almost similar system operates under the name *Jet-Prop*. The operation of the *Hydrodigger* is limited by the length (200 m) of the flexible hose attaching it to the mother vessel. The displacement of sand is claimed to be of the order of 1500 m^3/hr and somewhat less for cohesive sediments.

Sarcophaging of ammunition

The ammunition in the Baltic Sea is dispersed over a wide area, and it is impossible to cover all the dumping sites. In an emergency operation, sarcophaging should be limited to smaller sites where the CW ammunition is concentrated. For this purpose, a blanket of rock (stones and sand) 2 m high for a 250m x 250 m area would require 260,000 tons and cost approximately 7.1 million EUR.

Perhaps an alternative would be to excavate large holes at strategic points and to bury collected ammunition in these holes under a layer of rock. The holes could be dug by the deployment of the abovementioned ROV's equipped for either dredging, jetting or hydro-dynamic excavation"

An additional method is a grab system with lateral transport, operated from a surface vessel (**Fig. 22).** The large clamshell is operating hydraulically. Once the hole is ready, a working class ROV with manipulators is deployed to search the seabed and to lift and transport the ammunition underwater to the hole, where it is dumped. When the hole is full, it is covered with a sheet of rock. The ROV is powered from the support vessel, which has to follow the ROV during its movements underwater. A systematic approach is recommended, whereby the areas is subdivided in blocks, which are successively cleaned.

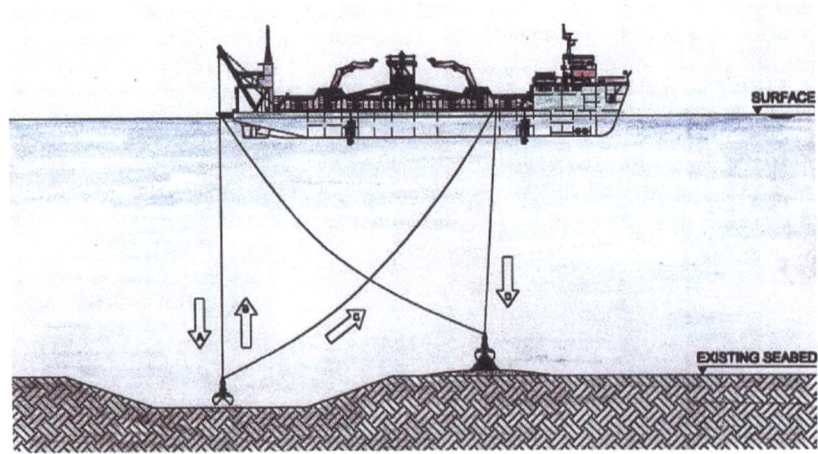

Fig. 22. Excavation and side depositing with a hydraulically operated clamshell system from a Dynamically Positioned (DP) vessel.

Disadvantages

Not all ammunition is easy to collect, since much of it will be covered with sediment and dispersed. There is also a real danger that intact ammunition may explode and that damaged ammunition, in particular the aircraft bombs, may start to leak and contaminate the ROV's, which would then be difficult to retrieve without contaminating the support vessel and its crew.

Other methods, i.e. **Fig. 23**, should be developed for such dangerous situations.

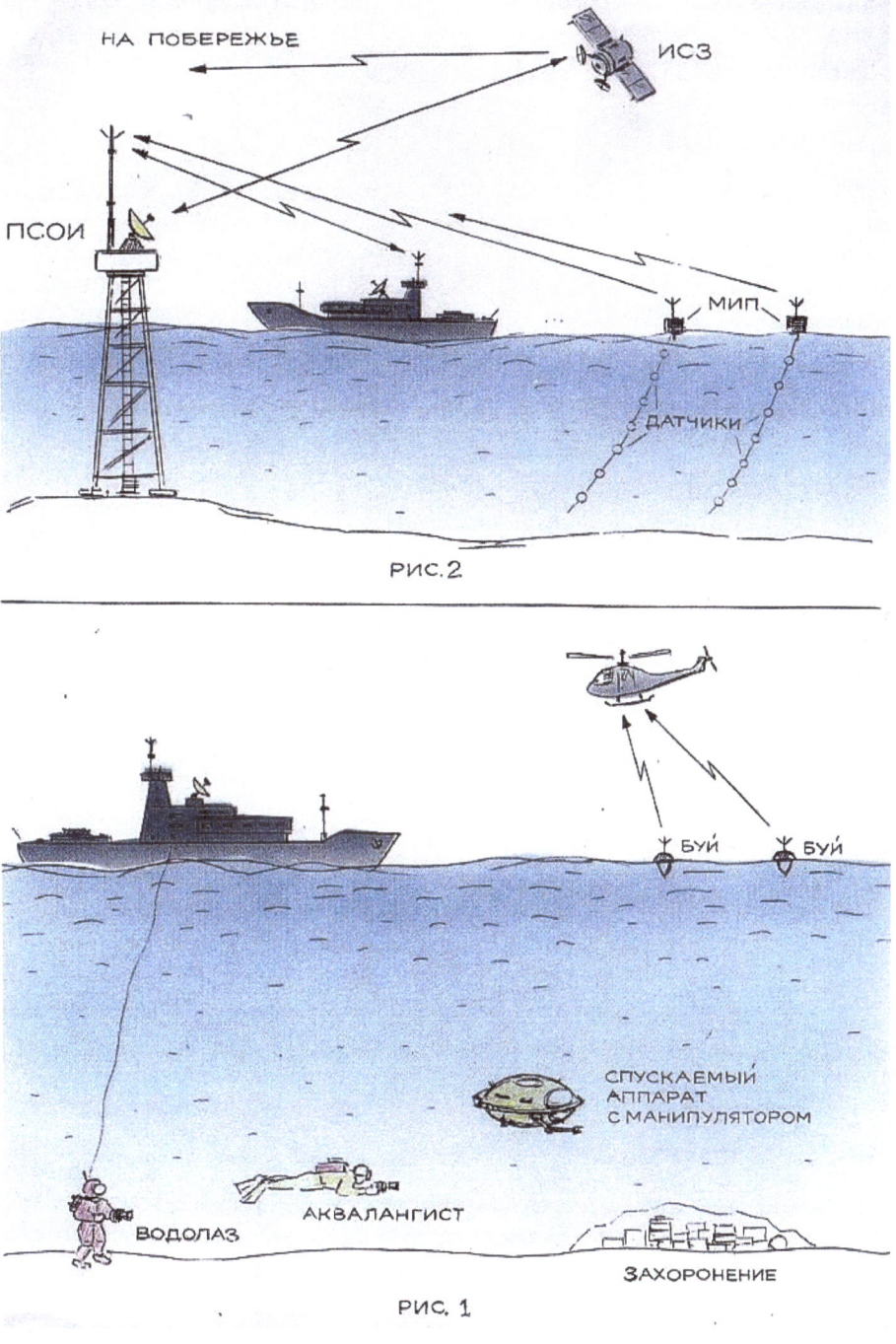

Fig. 23. Plans for searching of dumped ammunition (above) and their recuperation (below), as proposed by Kurganov and Morozov (1999).

5.1.4. Cost evaluation

Sarcophaging of wrecked ships

For large quantities of stone, the price would be in the order of 27.30 EUR (1999 price level) per ton. This unit price, considering the deposition of 20,000 tons per week, includes the following activities: At present (2014) an increase of 15% is not unrealistic.

*   Design procedures and documentation.
*   Pre-installation of geotechnical and geophysical surveys. ROV inspection of the sites.
*   Regular surveys during the sarcophaging, using the fall-pipe ROV.
*   Mobilisation and demobilisation of the stone-dumping vessel.
*   Loading of rock (stones, sand and silt) in a Norwegian harbour, travel to and from the site.
*   30% extra for mechanical and weather downtime
*   Insurance.

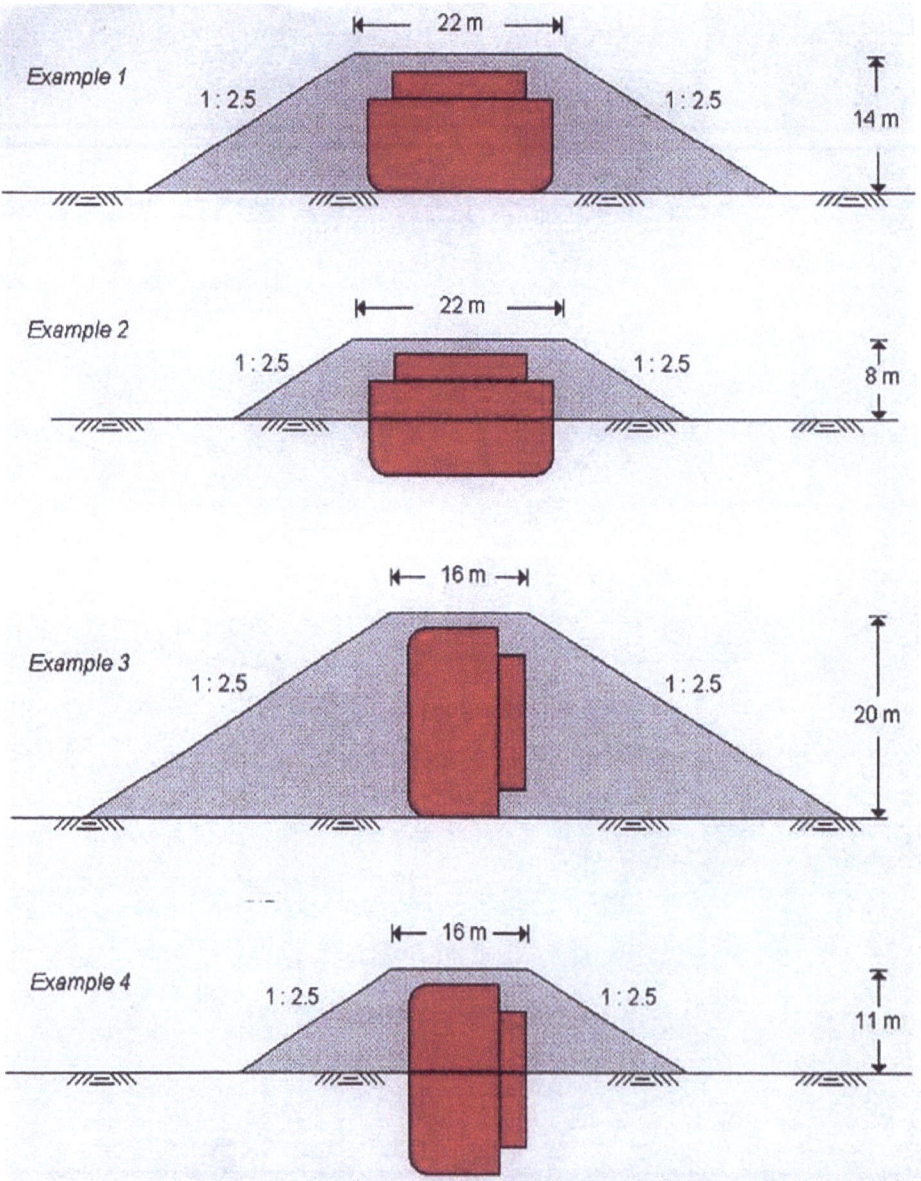

Fig. 24. Four typical sarcophage cross sections.

For a Liberty ship 140 m long and 18 m wide with a deck height of 12 metres, examples of four different situations can be taken (**Fig. 24**):

Example 1: The vessel has landed on its keel and is laying on the seabed. Assuming a 30 % loss of deposited stones and sand, the total volume would be 250,000 tons and the operation would cost 6.8 million EUR.

Example 2: The vessel has sunk into the seabed, but is still upright on its keel. The amount of material for sarcophaging would be less: 115,000 tons, and would cost 3.2 million EUR.

Example 3: The vessel is laying entirely on the seabed on its side: 550,000 tons would be required (15 million EUR).

Example 4: The vessel is laying on its side and has sunk into the seabed: 175,000 tons would be required (4.8 million EUR).

Application of the mentioned ROV and Hydrodigger

The daily rent of a working class ROV, such as presented in Fig. 18 is in the order of 9,000 EUR, 1999 prices, to which 20,000-23,000 EUR for the cost of the support vessel has to be added.

These prices for the *Hydrodigger* (Fig. 19) and the support vessel are: 14,000 and 25,000-27,000 EUR, respectively.

5.2. Methods to destroy leaked material

The Russian contribution to these problems of toxic substances leaking from chemical weapons dumped in the Baltic Sea, is in their participation along the following lines of research and technical development:

*   development of the general concept of the Baltic Sea decontamination (Fig. 22);
*   search for engineering solutions based on experience in the field of deep-water exploration;
*   creation of new highly-efficient, reliable and safe plasma-chemical reactors for under and above-water treatment of toxic substances.

5.2.1. In situ techniques

A.I.Mikulin

Research conducted by the Russian financial and industrial company "Ecotransenergomash", the designer-shipbuilders of St. Petersburg, Severodvinsk, Nizhny Novgorod and the designers of the Russian Airspace Complex has produced a promising method for transforming underwater chemical dumps by chemical, plasma-chemical and electro-discharge methods into soluble and insoluble components and for the transport of the soluble components in containers to a mother ship for treatment on-board.

Technical feasibility

1. Technically it is not complicated to erect a two-layered dome 50-70 metres high with a diameter of about 200 metres and consisting of an inflatable membrane and a stabilising concrete ring. Modern shipbuilding and airspace industries are highly developed and capable of producing durable hermetic domes of the necessary size.

2. The technology of processing under a dome is based on a similar kind of experience in surface conditions for extracting deposits of gold and uranium. Available chemical engineering facilities can assure the undertaking of such processes in closed volumes above the seabed.

3. Technical systems of pumping and transport are determined by the requirements for absolute ecological safety. In depths up to 100 m, such safety would be ensured by a special

pipeline system hermetically joined to a dome. Below 100 m, the employment of an underwater transport tanker would be required.

4. When processing chemical products on mother-ship-factories, the major problem is to produce sufficient power for the treatment of huge water masses, but however complex the techniques may be, the various types of plasma-chemical reactors developed within the last decade can cope with the technology.

Substantial studies by physicists and chemists should be conducted simultaneously with the design and creation of the entire complex so that the correct decisions are made.

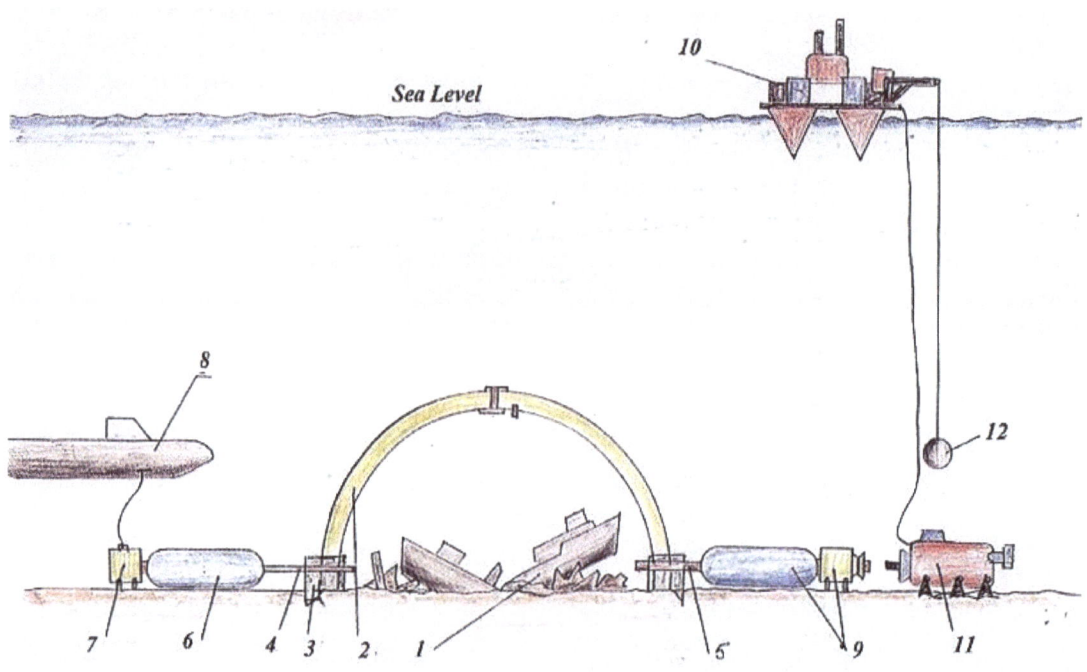

Fig. 25. Basic scheme of underwater treatment of leaked CW agents from dumped chemical ammunition

The concept is shown in **Fig. 25**. A two layer dome (2), is erected above a wrecked ship containing a CW cargo or above CW ammunition (1); an inflatable membrane is fixed on the seabed by a concrete ring (3). The dome (2) is equipped with two pipes (4) and (5), through which the input of chemical reagents and plasma-chemical or electro-discharge reactors and the output of products of underwater chemical reactions are effected. Pipe (4) is connected with underwater containers of chemical reactants(6) and a junction (7) is made with an underwater tanker (8), which ensures delivery and replenishment of the chemical reactants. Pipe (5) is similarly connected with a junction (9) for unloading liquid reaction products from under the dome (2) and their transfer to a mother ship (10) for further treatment on board. Depending on the depth, the transfer of liquid products may be carried out either through flexible pipelines connected with the junction (9) or through hermetic inflatable cylinders (12) filled by a special underwater tank-device (11).

Cost evaluation
It would take a year to set up an expert co-ordinating centre and to create working groups of technologists, designers, chemists and physicists to develop the concept and the technological scheme for the destruction of dumped chemical substances; the estimated cost would be $500,000.

Remark editor

Since the major threat is caused by lumps of mustard gas instead of contaminated water, the concept should focus on removing and destruction of these lumps either *in situ* or on board of the mother ship, perhaps in symbioses with the plans given in **Fig. 23**.

5.2.2. Incineration
E.K.Duursma and B.T.Surikov
Several sites are in operation under the control of the OPCW[80] concerning the incineration of CW stock piles on land.

Open-Pit burning
Since no other safe way of destruction was available,[81] this method was used just after WW-II and recently in Iraq to destroy 122-mm rockets, with the permission of the UNSCOM (United Nations Special Commission in Iraq). The CWC (Chemical Weapons Convention) does not however, allow open-pit burning.[82]

Incineration plants
Most CW agents, with the exception of sarin, are inflammable and incineration is practically 100%, if properly controlled.[83] In the US installation at the Johnston Atoll in the Pacific,[84] the CW agents are burnt in a counter-flow rotary device for 15 minutes at 538 $^{\circ}$C and in a subsequent after-burner for 1 second at 1200 $^{\circ}$C. The environmentally hazardous products that are formed after incineration are: nitrogen-, arsenic- and phosphorus oxides, hydrogen-chloride and chlorine. Chlorinated toxic dioxines may be formed if the temperature is between 180 and 400 $^{\circ}$C and when chlorine and reactive hydrocarbons are present. Therefore, during the destruction process, the period during which the incineration "off-gases" are heated should be minimised. Rapid cooling of the off-gases by water to 60 $^{\circ}$C and removal of hydrochloric acid and chlorine with sodium-carbonate solution and the final sorption of the gases with activated charcoal will ensure that no dioxines are released into the atmosphere.

5.2.3. Chemical methods
E.K.Duursma
The techniques to destroy CW agents by chemical means are in full development.[85] They have certain advantages over incineration since CW agents are converted into non-toxic compounds which are easier to dispose of.
Two-stage technologies are here mentioned:
*   *Organophosphorus CW agents*
A two-stage technology has been developed by GosNIIOKhT for organophosphorus CW agents.[86] This technique is characterised by its simplicity and meets Russian design criteria. The first stage is to react sarin for example with monoethanolamine at 100 $^{\circ}$C, resulting in a reaction mass 20,000 times less toxic. The second stage consists of binding the reaction

---

[80]OPCW (1998).

[81]Stock (1998), p. 81.

[82]Part IV (A) of the Verification Annex, para. 13.

[83]Stock (1998), p. 83.

[84]Stock (1998), p. 88-89.

[85]Chimiskyan (1998), 14-29. See also Greenpeace (1990).

[86]Sheluchenko and Utkin (1998), pp. 116-118 and Beletskaya (1998), pp. 107-108.

product (at 135 °C) with a mixture of bitumen and calcium hydroxide. This mixture is then heated to 200 °C at lowered pressure for one hour. Since sarin contains fluoride, these ions are bound during the bitumization into insoluble calcium-fluoride salts. The calcium hydroxide breaks up the phosphorus-organic binding, leading to the formation of an alcohol and a non-toxic calcium-phophorus compound.

*   *Mustard agents*

The most promising destruction option here seems to be a three-stage process of mustard hydrolysis, followed by a treatment with monoethanolamine and bitumization.[87] The hydrolysis takes place with calcium hydroxide and the next two stages take place at 100-110 °C at hourly intervals. The solid bitumen blocks can be safely buried as non-toxic waste.

Here new methods of neutralizing should be developed for treating mustardgas lumps in situ, using chemical methods which either rapid or slow react.

A natural way of decay has been observed by Medvedeva et al, (2009) by bacterial decomposition of so-called MGHP's (mustard gas hydrolysis products) along the reaction schemes as given below.

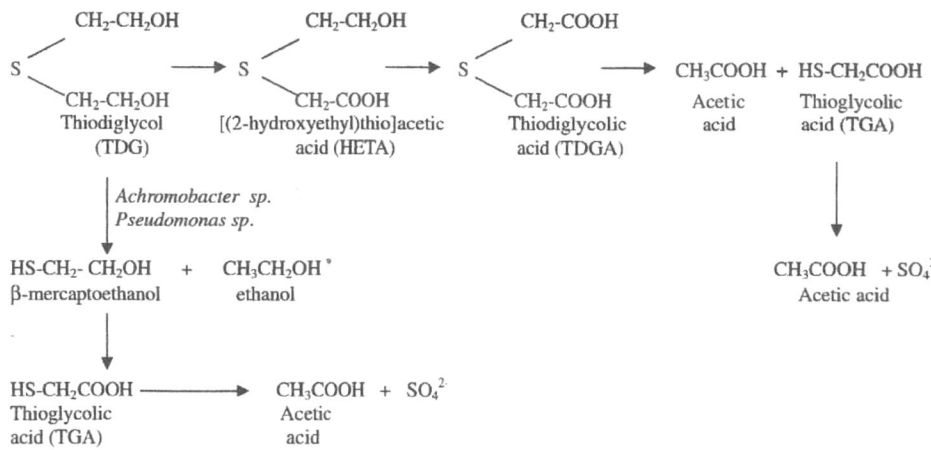

Their results suggest the potential of MGHP's biodegradation by naturally occurring populations of microorganisms in near-bottom water and sediments. It was not mentioned whether such biodegradation occurred with the original product, mustard gas.

5.2.4. Plasma-chemical techniques

I.A.Kossyi

Plasma-chemical techniques of incineration and decomposition by chemical reactions can, with certain advantages, also be used to destroy on sites dissolved and gaseous-phase toxic substances extracted from sea-dumped CW agents. Plasma-chemical devices are used for purification of exhaust gases, for destruction of ozone-depleting freon (CFCs), for extraction of ultra-pure silicon (silicon dioxide, silicon nitride) from gaseous-phase products, and for precipitation of diamond and diamond-like films, etching of metal and dielectric samples, etc. The thermo-non-equilibrium plasma-chemical reactor seems to present a quick, reliable and safe way to destroy CW agents and thus help to solve this pressing problem.

---

[87]Stock (1998), p. 91.

Five types of microwave plasma-chemical reactors can be distinguished:

1.A plasma-chemical reactor on the basis of a high-power pulse microwave generator (Fig. 25).

In a metal chamber the gas-flow containing the toxic substances, brought into gaseous phase, is exposed to microwave discharges. The gas outside the microwave beam remains close to room temperature whereas within the microwave beam, this temperature does not exceed 500 K (227 °C). At the same time the electron component reaches 50,000 K (49,727 °C) at an electron concentration of 10^{15} - 10^{16} cm^{-3}.

The efficiency of various kinds of plasma-chemical processes is dependent on time. Extrapolating from known investigations, the energy costs are about 1 kWatt.hour/kg, which means that a one-kilowatt unit will allow processing of about 24 kg of toxic substance per day.

The advantages of the method are:
* a high efficiency (approximately ten times higher than by incineration);
* the low temperature of the reactor;
* possibility of installation on a mother ship, in spite of its size (3x3x1m).

The disadvantage is:
* the cost of a microwave generator which ranges between $30,000 and $50,000.

Obviously, this microwave reactor can only be used on mother ships. In the scheme described in the previous section, once the products of decomposition of dumped chemical weapons have been recuperated as a contaminated sea-water or lumps of CW agents and transformed into the gas phase, the pulse microwave discharge reactor can execute the final purification.

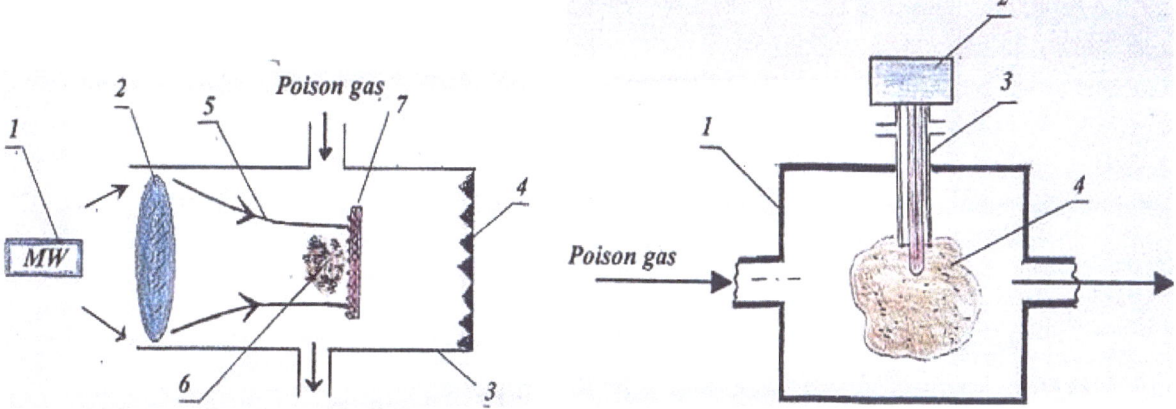

Fig. 26. Scheme of a reactor, based on a powerful pulse microwave generator. 1. microwave generator; 2. dielectric lens; 3. metallic chamber; 4. microwave absorber; 5. microwave beam; 6. microwave discharge; 7. initiator of discharge.

Fig. 27. Scheme of a reactor based on the domestic microwave generator. 1. metallic chamber; 2. magnetron; 3. coaxial wave-guide and initiator of the discharge; 4. microwave discharge

2. A plasma-chemical reactor on the basis of a consumer microwave generator

The version presented in **Fig. 26 & 27** is a typical 600 Magnetron of continuous action, commonly used in consumer microwave ovens. The device allows for microwave discharge at relatively high pressures in various gases and gas mixtures. When working with argon, it is possible to produce plasma at pressures reaching atmospheric value. When working with gas mixtures containing electronegative components (for example, CCl_4, $SiCl_4$, CF_2Cl_2, etc.), lower pressures are used. Low costs for the decomposition of chlorofluorocarbon molecules and for mixtures containing gaseous-phase toxic substances have been obtained.

The narrow interval of the working pressures is a drawback which may have an effect on the capacity of the device, but the extreme cheapness of the microwave generator and its small dimensions are the essential advantages.

3. A plasma-chemical reactor on the basis of a slipping surface discharge

This reactor (Fig. 27 and 28) differs from the former reactors in that it uses a device based on stimulating slipping surface discharges as in a plasma generator.

Its advantages are:
* simplicity and cheapness of production of discharge elements;
* possibility of producing flexible dischargers which create plasma of complex geometry (a ring, a spiral, a sphere, etc.);
* low costs of about 1-5 kW.hour/kg.

The reactor can be used on mother ships in the same way as the microwave discharge reactors and underwater vehicles, so its field of application may basically expand in comparison with microwave facilities.

4. A plasma-chemical combustion of toxic gases

The reactor based on plasma-chemical combustion of gaseous-phase toxic substances is similar to that represented in **Fig. 28 & 29**. The main difference from that described in the previous section is that the toxic product is mixed with hydrogen and oxygen in a non-detonating ratio. Simultaneous quick burning takes place in the whole volume of the chamber, the toxic products being virtually decomposed during one run (one high-voltage pulse).

Energy costs are extremely low (about 0.01-1.0 Watt.hour/kg). The method is applicable to all kind of CW agents found dumped in the Baltic Sea, and can be used on mother ships.

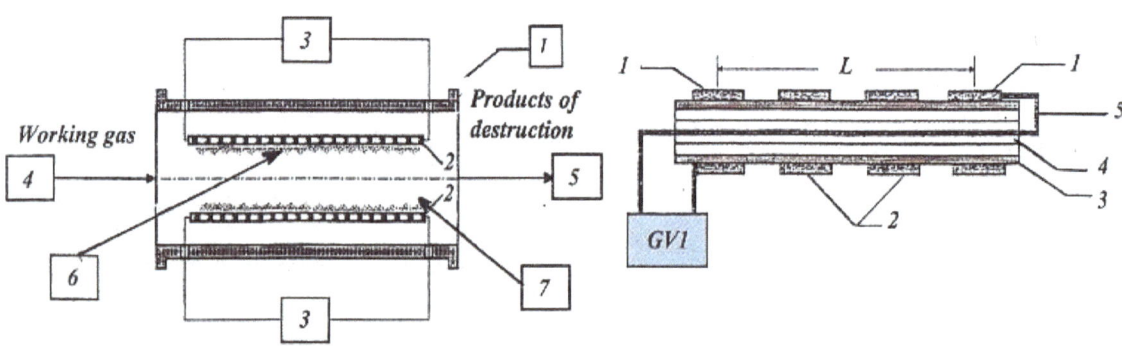

Fig. 28. Scheme of a reactor based on a slipping surface discharge. 1. metallic cylinder (tube); 2. slipping surface discharger; 3. high-voltage pulses supply; 4. source of treated gaseous mixture; 5. treated gas absorber; 6,7. gas-discharge plasma.

Fig. 29. Slipping surface discharger. GVI-high-voltage pulses supply. 1. main electrode; 2. intermediate electrodes; 3. ceramic tube; 4. insulator; 5. back-current rod.

5. Electro-discharge purification of water from dissolved toxic substances

Plasma-chemical research is also directed towards the purification of water in Russia, the USA, South Korea and England.[88] The concept of the reactor is represented in **Fig. 30**

---

[88]Shmelev, Evtyukhin and Che (1996) and Goryachev, Rutberg and Fedyukovich (1998).

chamber (pipe), through which the processed fluid flows, contains an electro-discharge device of a special design,[89] containing a system of local electrical discharges in the water medium. The plasma formation, generated on the surface of the discharge device, removes various chemical impurities from the water with a high efficiency.

The purification ability of the discharge device is dependent on the synergetic action of various factors:

* high-power ultra-violet radiation of electro-discharge plasma;
* strong sound and shock waves exited by the discharge;
* production of chemically active substances (i.e. O_3) and radicals (i.e. OH);
* initiation of cavitation phenomena.

These factors add up to a low energy cost of purification of 1-10 Watt.hour/litre, or a cleaning capacity of 1 m^3/hour/kWatt. The method could in principle be used on a mother ship, but has not yet been tested for CW agents.

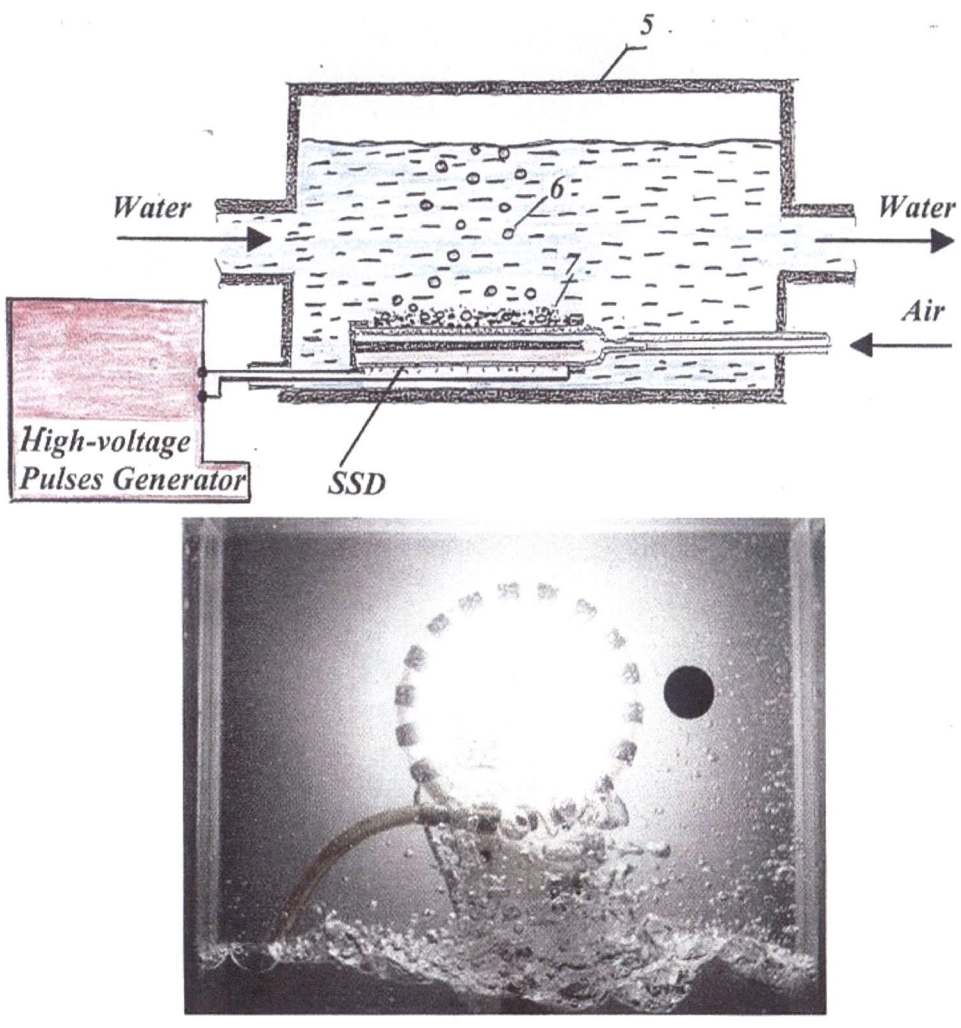

Fig.30. Above: Scheme of a reactor for cleaning contaminated water containing poisonous substances. 1. metallic chamber; 2. air bubbles; 3. electric discharge plasma. 4. SSD (slipping surface discharger). **Below:** Photograph of a laboratory installation with circular plasma electro-discharge purification.

---

[89]Created and tested at GPI (General Physics Institute of the Russian Academy of Sciences).

Cost evaluation

In order to set up a programme of plasma-chemical reactors of detoxification of CW agents dumped in the Baltic Sea it would be necessary to conduct:

1. Experiments on the efficiency of microwave discharges, of slipping surface discharges and of plasma-chemical combustion: execution time: 1.5 years; estimated costs: $400,000

2. Experiments on the electro-discharge utilisation of removing toxic substances dissolved in sea water. Execution time: 1 year: estimated costs: $300,000

3. Development of designs of plasma-chemical modules for surface and underwater utilisation of toxic substances: execution time: 8 months; Estimated costs: $100,000

4. Production, mounting and start-up of demonstration reactors for plasma-chemical utilisation of toxic substances: execution time: 1 year; Estimated costs: $500,000.

6. OCEANOGRAPHIC DATA *(Duursma)*

6.1. Maps of bottom currents in the Baltic

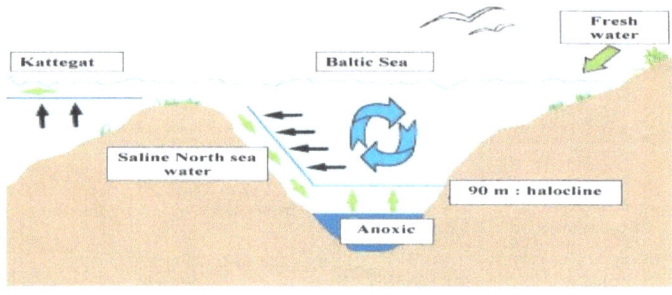

Fig. 31. Rough scheme of the water exchange in the Baltic Sea.

The Baltic Sea is by its nature a kind of inland sea, with out- and inlets through the streets of Denmark **(Fig. 31).** .Its sea water ranges from fresh water close to Finland to North Sea saline water entering over the bottom from the Kattegat. This causes in the Baltic proper a layered system, with the higher salinities along the bottom. In the deep regions anaerobic conditions prevail.

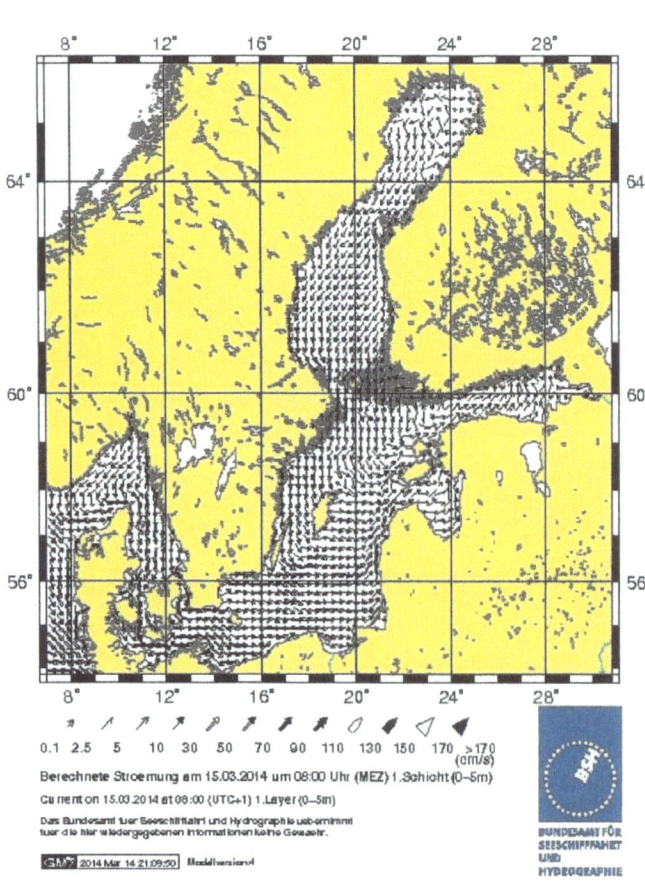

Fig. 32. Surface tidal currents example as available from the Bundesamt fur Seeschiffahrt und Hydrography, Hamburg

Due to this circulation, the bottom currents, which can transport lumps of mustard gas have a trend inwards, however not with clear directions, depending on the wind-driven surface and tidal currents. These **(Fig. 32)** range with currents of 2.5 to 150 cm/sec, which is for the weakest current 2.15 km/day. If lumps travel only 10% of this value, they move 215 m/day or 78 km/year.

This calculation is, however, only a hypothetical one, and it becomes clear that rest current data and lump/rest current effects should be available for modelling the lump distribution in the future.

Our proposal is therefore that **pilot studies** are made in hydrological laboratories to determine the basic items for modelling the lump distribution. A task for he CHEMSEA[90] group of Baltic Sea institutes and for example the Hydraulic Laboratory DELTARES, Delft, The Netherlands.

6.2. Maps of bottom currents in the Skagerak

The distribution of mustardgas lumps from the shipwrecks in the Skagerak should equally be studied. The same pilotplan studies are proposed where the data can be applied with the hydrographical current data of the seas north of Denmark. As both **Fig. 33** and **34** show, the tidal currents have a reststream north along the west coast of Denmark, towards the Swedish coast, bending than along the Norwegian coast northwards.

Although part of th Skagerak water will supply the deep saline water in the Baltic, the major waters turn around the Norwegian coast.

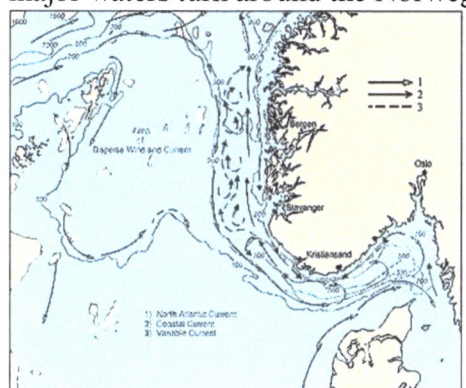

Fig. 33. Current directions in the Skagerak and Kattegat. [91]

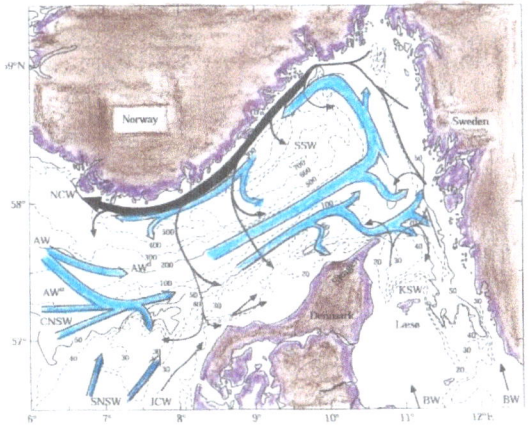

Fig. 34. General circulation of the watermasses in the Skagerak. Black arrows represent the rest surface current, the blue arrows represent the rest subsurface currents, taking tidal movements into account. [92]

[90] CHEMSEA, Results from the CHEMSEA project –Chemical munitions Search and Assessment. ISBN: 978-83-936609-1-9

[91] Sailing directions 2013 Skagerak and Kattegat Nat. Gespace Intel. Agency AGENCY Springfield, Virginia 14th ed., 193.

[92] Danielssen, D. S., EDLER, L., Fonselius, S., Hernroth, L., Ostrowski, M., Svendsen, E., and Talpsepp, L., 1997. Oceanographic variability in the Skagerak and the northen Kattegat, May-June, 1990. ICES J., Mar. Sc., 54, 73-773.

6.3. Pilot studies

6.3.1. Displacement properties of lumps.

After dumping chemical munitions in the Baltic Sea, the Admiral of the Soviet Marine, who was in the hospital with Major-General Boris Surikov, warned him on the corrosion of in particular the bombs containing mustardgas. Others predicted holes within 50 years. It is now 67 years later, and it becomes more and more urgent that at least knowledge is achieved on the mustardgas lump transport along the bottom.

Pilot studies can be made by hydraulic laboratories, such as shown in **Fig. 35** or at smaller scales. Test can be done with harmless material similar as the mustardgas lumps for different current velocities and slopes.

Fig. 35. Stream installation of the Delft Hydraulic laboratory Deltares. Photo obtained by courtesy of Deltares.

Once the data obtained, modelling should be possible, using the data such as given in **Fig's. 32 to 34** and information on the slope of the bottom.

7. INCIDENTS AND PROTECTION

7.1. Human encounters with dumped chemical weapons
E.K.Duursma

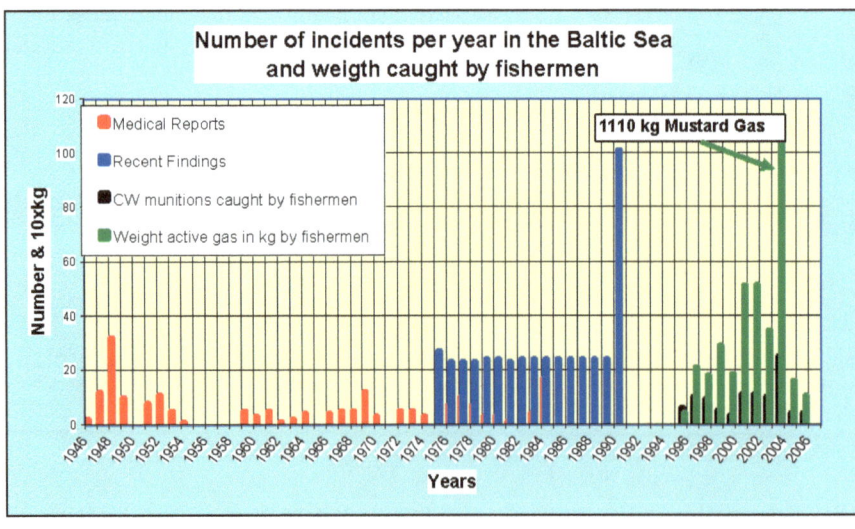

Fig. 36. Number of incidents with CW ammunition as recorded by medical reports and recent registered findings (Denmark) Updated until 2005

Since 1946, human contact with dumped chemical weapons (**Fig. 36**) has certainly occurred, mainly by fishermen bottom-trawling in risk areas of the Baltic Sea and ignoring warnings to avoid dumping sites. Between 1946 and 1984, 197 cases of people suffering from mustard gas exposure were reported and a total of 171 patients were treated; 26 of these were actually admitted to hospital.[93]

The Helsinki Commission has yet (2013) published more clear figures, although several Baltic States, such as Germany have no legislation on the obligation to report findings of CW material. For Denmark, where fishermen are compensated for retrieval of captured CW material, 439 were recorded between 1976 and 1992 (101 of these in 1990[94]). The economical loss is not negligible as the data plotted in **Fig. 37**[95] demonstrate.

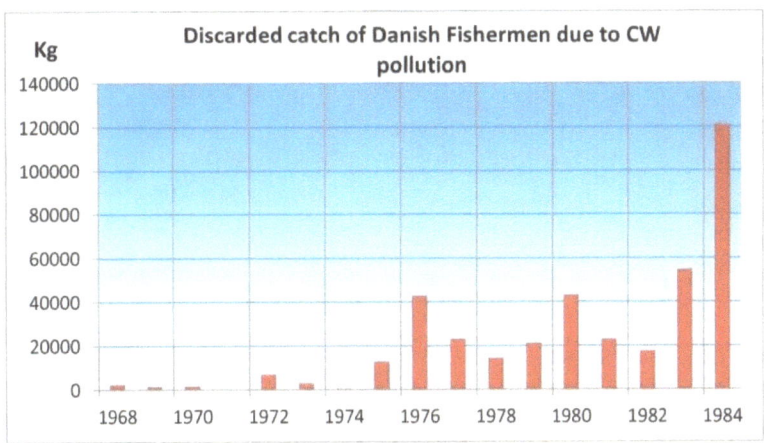

Fig. 37. Discarded fish catch between 1868 and 1984, as reported by Helcom (2013).

After 1984 and from 1994 on, also the recovered weight of mustard gas during incidents were evaluated. They are given in **Fig. 38.**

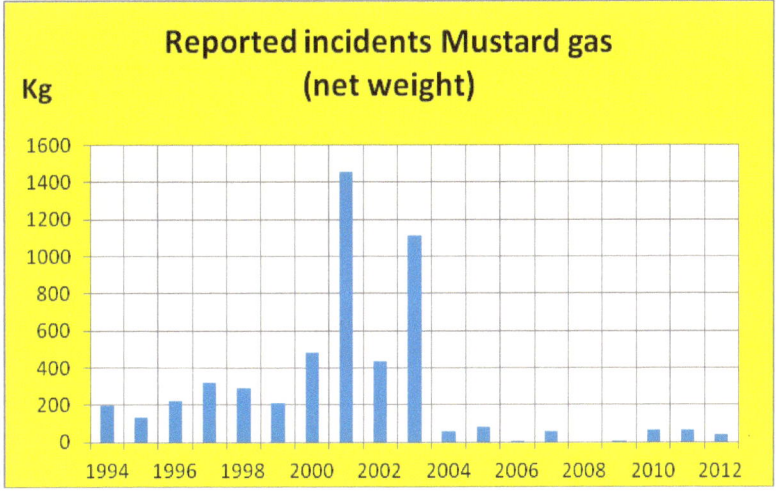

Fig. 38. Weight of recovered mustard gas during incidents in the Ba;tic Sea (Helcom, 2013).

Findings in the Baltic Sea often occur outside the official dumping sites (**Fig. 39**)96 and 123 people were burnt between 1955 and 1970 by CW agents on Polish beaches, where high seas had carried barrels of mustard gas. The most serious incident was in 1995 when 120 children were playing near an eroded barrel. The first symptoms of skin burning and severe eye injuries appeared after only 30 minutes. Another accident occurred on 28 January 1997 when the fishing vessel KOL-158 caught 20 kg of a strange clay-like material in the central

---

[93]MEDEA (1997).
[94]Theobald and Rühl (1993).
[95]MEDEA (1997).
[96]Korzeniewski (1994), p. 93-96 and Korzeniewski (1999).

part of the Baltic Sea. The material was immediately thrown overboard, and the vessel returned to its home port, but the crew suffered the next morning from the first symptoms of burning.

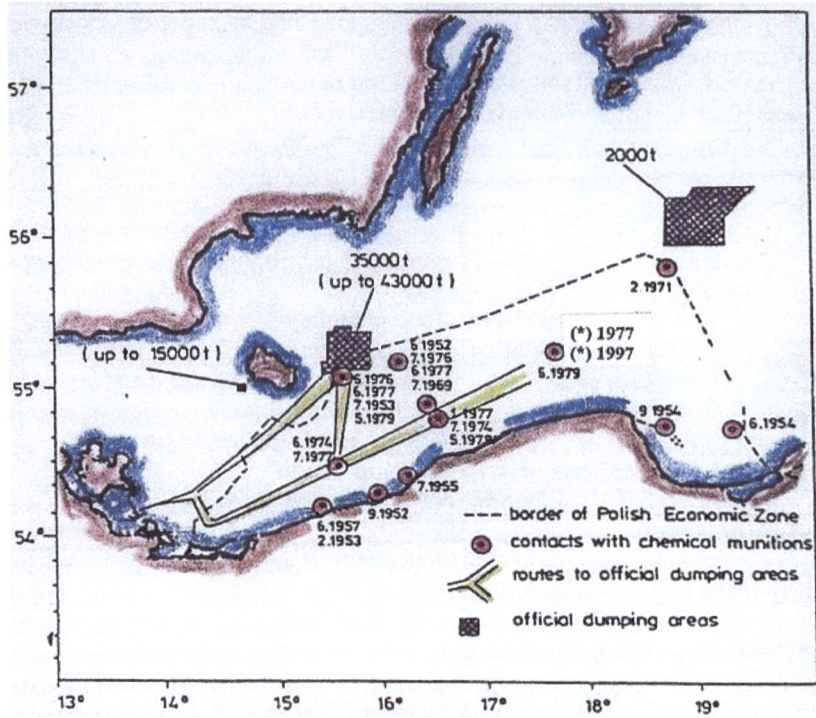

Fig. 39. Dates and locations of contacts with chemical munitions as reported by Korzeniewski (1994) in the Polish Economic Zone. This figure is corrected by Korzeniewski (1999), adding 7 incidents with CW ammunition in 1967 east of Bornholm, 7 incidents in 1969 south of Bornholm and 1 incident each in 1977 and 1997 as shown in figure by (*).

In the Dutch legislation, the authorities allow the destruction of captured conventional munitions by competent sections of the Coast Guard. The total annual budget for this is limited to Hfl. 100,000.- (Hfl. 1 = 0.45 EUR) with a compensation for the first captured munitions (mine or bomb) of Hfl. 400 and for subsequent captures within the same week of Hfl. 100.[97] This amount is not usually enough to compensate for the loss of one or two days fishing. It is therefore not unusual that captured munitions are thrown overboard instead of brought on land.

7.2. Symptoms and first-aid
E.K.Duursma
Lethal concentrations, Symptoms and first-aid treatment
On the basis of information obtained by the courtesy of Commandant Christian Chevallier and Adjudant Jean-Marc Decaunes of the Sapeurs-Pompiers de Monaco and from the Royal British Navy98 through Rear Admiral Gilbert de Cock, a table summarising the mechanisms of action, lethal concentrations, symptoms and first-aid treatments have been prepared for the five classes of CW agents dumped in the Baltic sea and Skagerak (Table IA and IB).

[97]Bijstands- en bijdrageregeling opgeviste explosieven. Kustwachtcentrum, IJmuiden, Netherlands.
98HMSO (1987).

Table IA: Type of warfare gases, their mechanisms of action and lethal concentrations (See also Table II in Annex I.) Cl-acetophenone as tear gas is not included in this table.

Type of CW agents	Mechanism of action Lethal concentration
Nerve agents (Organo-phosphorus compounds)	Inhibition of the enzyme acetylcholinesterase causing poisoning by excessive accumulation of acetylcholine.
Tabun, GA	400 mg-min/m^3
Sarin, GB	100 mg-min/m^3
Soman, GD	50 mg-min/m^3
Blood agents	Cyanide forms reversible complex with respiratory cytochrome oxidase enzyme system. In particular is the central nervous system of the respiratory centre susceptible and failure causes death.
Hydrocyanic acid	5000 mg.min/m^3
Cyanogen-chloride	11000 mg.min/m^3
Lung-damaging (choking) agents	Increase of permeability of the alveolar capillaries with resulting pulmonary oedemia, interfering with gas exchange, leading to anoxia.
Phosgene, CG & Diphosgene, DP	3200 mg-min/m^3
Chlorine	20000 mg-min/m^3
Vesicant (blister) agents	Powerful alkylating action on DNA, and amino, thiol, carboxyl, hydroxyl and phosphate groups in cells.
N-Mustard (HN, N-Lost)	4500 mg-min/m^3
S-Mustard (Yperite, H, S-Lost)	1500 mg-min/m^3 (carcinogenic)
Lewisite, L (arsenical vesicant)	1300 mg-min/m^3
Nose and throat irritant agents	Blocking of enzyme action in cells through binding of arsenic by -SH-groups of these enzymes
Adamsite, DM (vomiting agent)	15000 mg-min/m^3
Clark I, DA	15000 mg-min/m^3
Clark II, DC	10000 mg-min/m^3 (all: 0.0037 mg As/l drinking water criteria LC$_{50}$ Fish As: 0.9-2 mg/l)

48

Table IB: Their symptoms and first aid treatments.

Symptoms	First aid treatments
By inhalation symptoms appear rapidly: Tightness of chest, rhinorrhoea, salivation, miosis with dimming of vision, difficulty in accommodation, frontal headache, transpiration and convulsions, cardiac arrest.	Use decontamination kit. Cleaning of eyes , skin, and clothes; slow intraveinal injection of atropine sulphate (2 mg AS) and pralidoxime (500 mg); swallowing a 5 mg tablet of diazepam; artificial respiration. Each half hour injection of 1 mg AS until slowing down of bronchial secretion.
Symptoms appear very rapidly on central nervous system. Powerful respiration and violent convulsions occur within 30 seconds and cessation of respiration within 1 minute. At lower concentrations vertigo, nausea and headache, convulsions and coma.	Immobilising, keeping warm, artificial respiration. Treatment: with dicobalt edetate (300 mg/20 ml) called Kelocyanor, but only in severe cases (is toxic to the liver and kidneys), to be slowly followed with intravenous dose Sodium thiosulphate (25 g in a 50% solution).
Latent period may last 30 minutes or more. Respiratory problems pulmonary oedema, dyspnoea, cyanosis, vomiting, convulsions.	Keeping warm and at rest, rinsing of eyes, treatment as for shock and bronchia-pneumonia with codeine phosphate (30-60 mg). Steroid inhaler can be life-saving, with an initial dose five times higher than for asthma. Also oxygenotherapy.
Possible latent period, redness of skin, eye troubles, diarrhoea, convulsions, fever, headache. Skin: blistering, necrosis extends into dermis. Respiratory system: inflammation followed by necrosis and pulmonary oedema. By swallowing: along the alimentary tract oesophageal and gastric mucosa, causing necrosis and perforation.	Keeping warm, rapidly rinsing eyes with plenty of water or 2% sodium bicarbonate. Treatment as for heavy-degree burns. Decontaminating skin and clothes by cleaning with petrol or oil, later with warm soap water. (Due to low solubility in water and higher solubility in non-polar liquids such as petrol and oil - see Annex I -; also oxidisable with hypochlorite).
Immediate pain in eyes, skin and respiratory system. Erythema, vesication and eye injury develops faster than with S-mustard. Effects being severe within 4-8 hours. Coloured urine, bleu lips, haemorrhage, tiredness, followed by systematic arsenic poisoning.	As for S-mustard casualties. In addition local treatment for eyes and skin with dimercaprol (BAL) 5% solution in arachis oil with benzyl benzoate. Can be used for intra muscular injection of 2.5 mg/kg body weight deep into the buttocks every 4 hours followed by 4x/day for 2 days, than 2x/day for 10 days.
Intense sneezing and couching, respiratory problems, head ache, dizziness. Later poisoning by arsenic compounds of liver, kidneys and red blood cells.	Inhalation of chlorine in weak concentration. Decontamination with hypochlorite, chloramine or permanganete solution.

HELCOM CHEMU:

In the final report of *ad hoc* Working Group on Dumped Chemical Munitions (HELCOM CHEMU) to the 16th Meeting of the Helsinki Commission, Attachment 3,[99] preventive measures and first-aid guidelines are given to the contracting parties in order to elaborate National Guidelines for fishermen on how to deal with caught chemical munitions. These guidelines can be summarised as follows:

What to do First
* Turn the vessel immediately to keep the crew up-wind. Close doors and stop the ventilation system.
* When on a beach, stay up-wind of the suspected item.
* Start immediate decontamination of people, even if no adverse effects are felt at first.
* Contact port authorities for instructions.

First-aid equipment
* One "gas box" for each crew member should be available on board, containing: 5 tongue spatulas; 4 packets of absorbent cotton; 3 bottles "Gas-decontamination liquid"; 3 powder sprays, "Anti-gas powder"; 1 bottle "Anti-phosphorus liquid"; 10 atropine/oxime automatic injectors; 1 instruction leaflet. Protection clothing, such as presented in Fig. 40.

Fig. 40. Protective clothing as recommended for decontamination of a contaminated fishing vessel. Illustration reproduced by courtesy of MATISEC, P.O.Box 26, 38080 St. Alban de Roche, France.

These instructions are further developed in the most recent Helcom report (Helcom, 2013), which can be obtained online from the site of Helcom at Helsinki.[100]

---

[99]HELCOM CHEMU (1995), p. 11-22. See Table I and also Annex I, concerning chemical properties of CW agents.

[100] HELCOM 2013 Chemical Munitions Dumped in the Baltic Sea. Report of the *ad hoc* Expert Group to Update and Review the Existing Information on Dumped Chemical Munitions in the Baltic Sea (HELCOM MUNI) BSEP) No. 142 Baltic Sea Environment Proceeding pages 128.

7.3. Long-term effects

E.K.Duursma

Few accessible documents exist on the long-term somatic and genetic effects of any kind of poison, but it is well-known that, once intoxicated, people may suffer for very long periods from vague to serious complaints, often difficult to diagnose. Golf-war syndrome is an example and also the intoxication following the EL-AL 747 accident in Amsterdam.[101] Similar cases may also occur after bacterial food intoxication, causing persistent allergy or sensitivity toward additives in frozen and canned food.[102]

Survivors from gas attacks during WW-I tell how they have suffered from "weak lungs" most or all of their lives. Charlotte Auerbach (1899-1994) was a world authority on mustard gas effects. In 1940, Professor A.J.Clark of the University of Edinburgh, asked her to discuss possible effects of mustard gas on gene mutation. Prof. Clark was impressed by the long-lasting effects of mustard gas on human cells: wounds were slow to heal and liable to open up again later; ophthalmologists in 1939 were still treating ulcers of the cornea produced by exposure to mustard gas in WW-I. These long-lasting effects seemed similar to X-rays effects. Thus it occurred to him that mustard gas, like X-rays, whose mutant effects were known, might also alter genetic material in the cell nuclei.[103] Charlotte Auerbach made most of her genetic studies with *Drosophila* flies[104] and found evidence not only for delayed mutations but also for the reproduction of these mutations. This latter phenomenon has remained controversial[105] and was "very much a puzzle" for her.

The greatest danger from sub-lethal intoxication by mustard gas may well be from genetic damage inherited by descendants. Although proved for *Drosophila* flies, the genetic damage depends on dose-response relationships and the effective filter in the embryogenesis. No documents have been found describing malformations of children whose parents survived the gas attacks of WW-I, or for A-bomb survivors in Japan[106] who received high levels of radiation. Adequate information on CW intoxication is difficult to obtain from modern geneticists[107] who work mainly on other topics. However the threat is there (**Fig.41 and 42).**

7.4. Ecological effects

E.K.Duursma and B.T.Surikov

The threat of dissolved CW agents to the Baltic marine environment itself can be eliminated according to information available to the Helsinki Commission. However, high levels of sparingly soluble clark, adamsite or viscous mustard gas can occur in the sediments in the immediate vicinity of dumped munitions and reports on the detrimental effects in the marine environment due to warfare agents have been recorded.[108]

---

[101]Telegraaf, 5 March 1999: Victims show auto-immune illnesses where their own defence system acts against their own body. Information given by H. Plokker, Chief Inspector for Human Health for the Governmental Hearing Commission.

[102]Own experiences.

[103]Beale (1995); Tarasov and Kalilina (1992).

[104]Auerbach (1947), Auerbach and Robson (1947), Auerbach and Moser (1950), Nasrat, Kaplan and Auerbach (1954).

[105]Kilbey (1995).

[106]Prof.dr.Peter Herrlich, Research Centrum Karlsruhe, personal communication.

[107]Prof.dr. Erhard Geiszler, Max-Delbrück-Centrum, Berlin, personal communication, 4.3.99.

[108]HELCOM CHEMU (1994), P. 31.

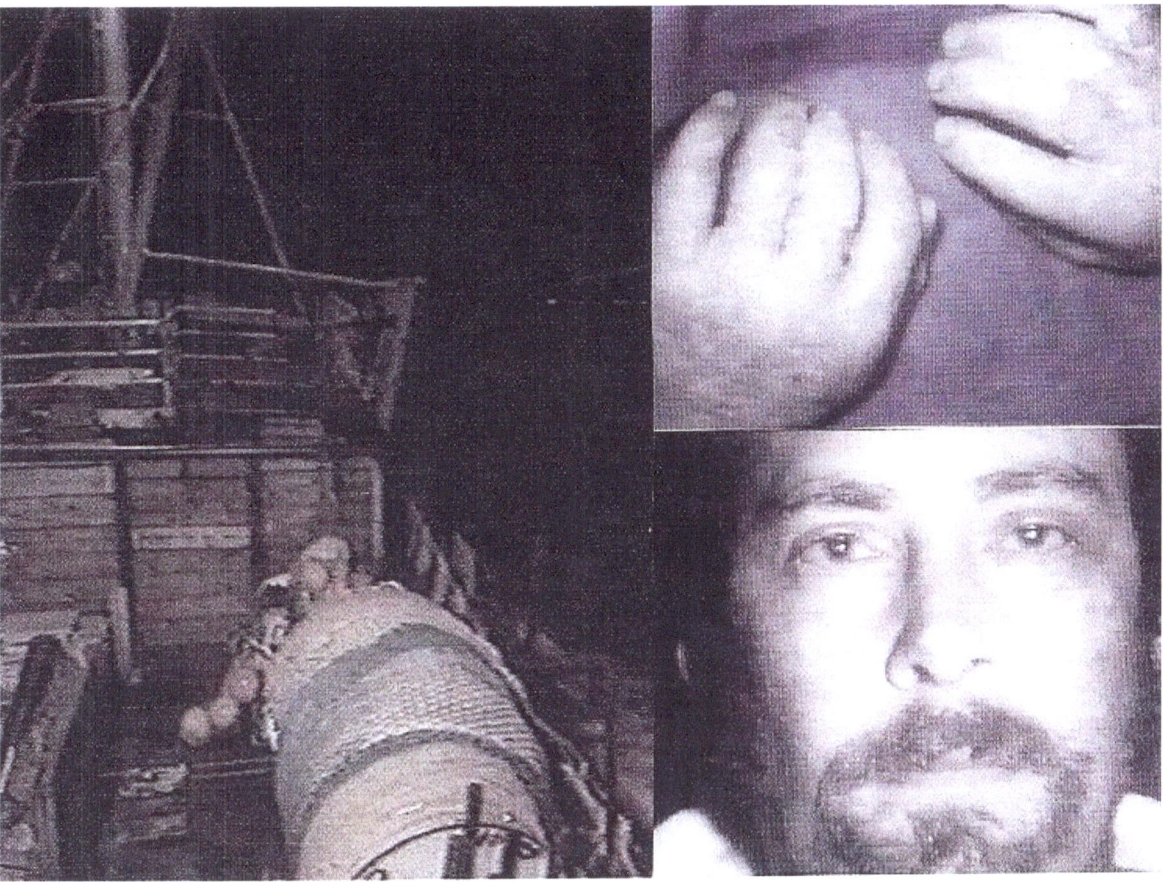

Fig.41.: A gas-bomb on board of fishing vessel!!!, A crew member is suffering from pain in the eyes, longs and skin. His hands are swollen and painful. Result of mustard gas intoxication. (Photographs reproduced with permission from the Creative Ass. Film Program - the XXth Century, Moscow).

The ecological catastrophe on the Letnii Coast of the White Sea's Dvina Gulf in May 1990, where 4 - 20 million starfish *Asterias Rubens* died, was probably due to CW agent intoxication. On 06/10/90, a girl who was playing with starfish died. Following another catastrophe in 1979, in which a mass death of bottom-dwelling fish was noted, official data confirmed that 700 aircraft bombs and over 5 tons of mustard gas-lewisite mixture in 31 iron barrels were dumped109 in the vicinity. There are at least ten hypotheses as to the cause of this disaster, and an official report by the Arkhangelsk Fishery Complex indicates that repeated tests showed traces of yperite (S-mustard) in samples of starfish, herring, mussels, seaweed, whitefish, flounder and navaga in the period May 23 1990 to June 7 1990; later, however, all samples were negative.110

---

[109]Yufit, Miskevich and Shtemberg (1995).
[110]Alimov and Khlebovich (1990).

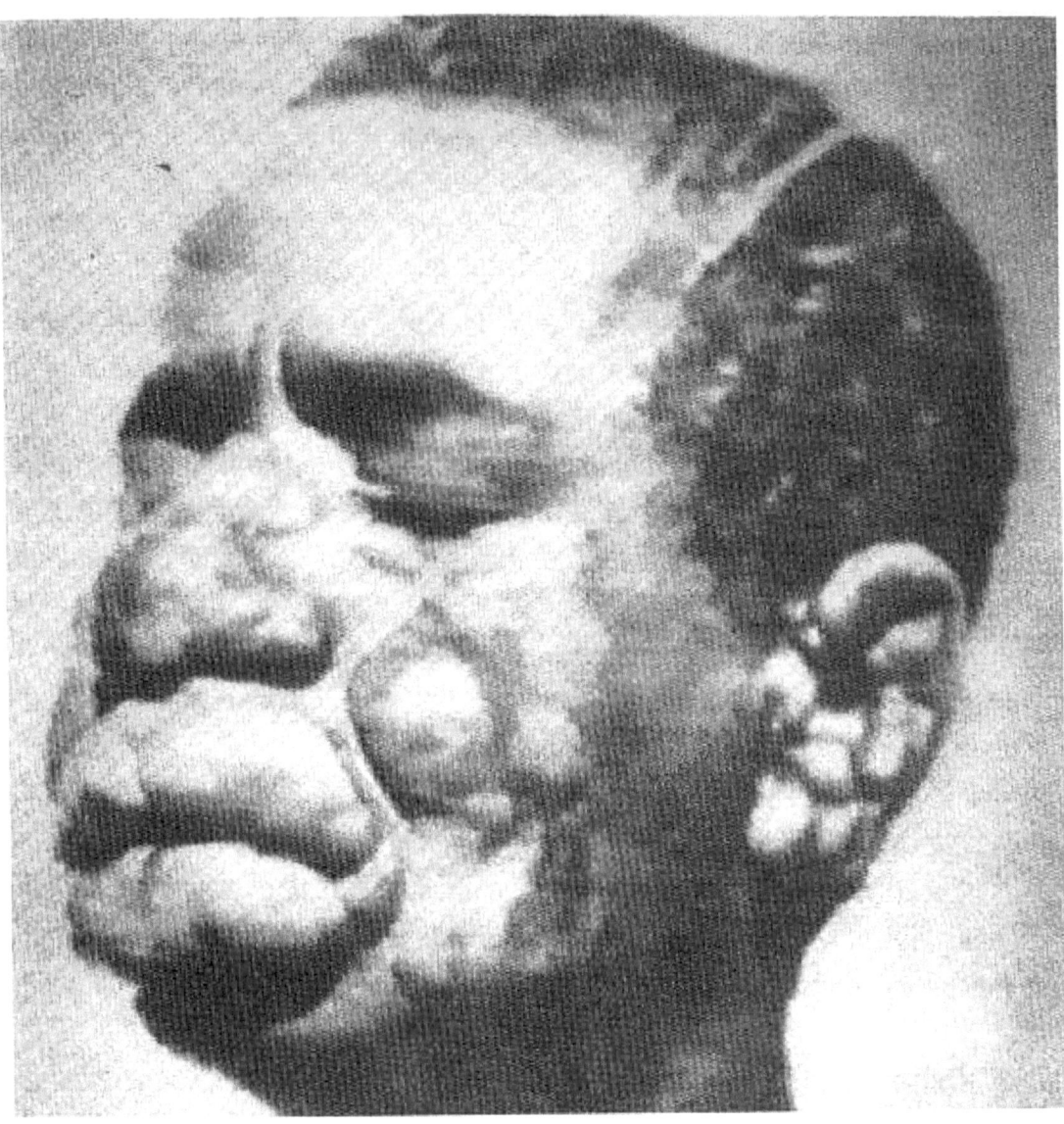

Fig. 42. Victim, blistered by mustard gas (Yperite) following an attack by the Italian army during their invasion in Abyssinia (Ethyopia). (Photograph reproduced with permission from the Creative Ass. Film Program - the XXth Century, Moscow).

Some CW agents, such as S-mustard and lewisite, have a higher solubility in lipids than in water (see KOW Annex I, Table II-B), and can accumulate in cells from a dissolved state in sea water.111 This does not mean that these products necessarily accumulate in the food chain. The determining factor is the ratio between their concentrations in water and in lipids, although intake may also occur from food. As far as mustard gas is concerned, and due its instability in a dissolved form, the most serious contamination is through contact with the lumps on the sea bottom, which is why so many starfish were killed in the Dvina Gulf.

---

[111]Duursma and Carroll (1996).

7.5. Insurance against incidents

Is it possible to insure sailors, fishermen and seaside visitors against accidents with poisonous CW agents? This question can and should be posed, since there is a population risk of contamination by CW agents in the Baltic Sea and Skagerak region.112

The question was put to two insurance companies:

J.H.Holsboer, **ING, Amsterdam.**113 The answer is that the likelihood of obtaining specific insurance cover against such risks is extremely small. Insurance for an acceptable premium is only possible if the risk is spread over a large number of insured people and it is doubtful whether the critical amount - *book of business* - could be attained. Furthermore the risk is difficult to quantify and there is a distinct danger of anti-selection, meaning that only people with significant exposure to the risk would be prepared to take out insurance. This is unattractive for insurers.

Although it is unlikely that specific insurance cover would be available there may be cover under policies of a **general** character, such as health, disablement and accident insurance and (in the case of seaside visitors) travel insurance.

However, under such an insurance, damage caused by CW agents might be excluded, because of the ⬚war risks⬚ exclusion which is to be found in many insurance policies. Under the Netherlands law,114 fixed premium insurers (as opposed to mutual insurance companies) are not allowed to cover war risks. Whether damage caused by CW agents can be classified as damage caused by " war risks " is open to debate and depends on the exact wording of the policy.

R.C.Seward, **The Britannia Steam Ship Insurance Association Limited.**115 Ship owners, including fishermen can insure against war-risks and this would include damage occasioned by accidental fouling of old bombs. This coverage can include the risk of injury to their crews. Tourists visiting the area, and insured by normal personal accident policies or travel insurance, would find it very difficult to arrange a special insurance without involving the responsibility of the governments concerned.

112This question was also discussed with Mr. Klaas Reinigert, Coyuro Management b.v., Salvage, Transport & Heavy Lift Consultants, Vlaardingen, Netherlands. He agreed with the reply given by Mr. Holsboer and suggested a second opinion by Mr. Seward, both replies given here.

113Holsboer, J.H., member of the Board of the ING GROUP, Amsterdam. Information obtained by letter of 29 September 1998 and adjusted by fax of 22 March 1999.

114Art. 64, par.2, Law on the Supervision of Insurance Companies (Netherlands).

115Seward, R.C., Tindall Riley (Marine) Ltd of the Britannia Steam Ship Insurance Association Ltd, London. Information obtained by letter of 2 March 1999.

8. BIBLIOGRAPHY
(Cited and consulted documents)
E.K.Duursma

Alimov, A. and Khlebovich, V. 1990. What has happened in the White Sea. PRAVDA, June 6, 1990.

Anonymous, 1992. Protocol of a meeting on the problem of dumped CW agents in the Baltic Sea. 2 pp., provided by Major General B.T.Surikov.

Anonymous, 1992. Final Act Resolutions on Dumped CW agents after WW-II. Diplomatic Conference, Helsinki, Finland, 34 pp. Document (in Russian) provided by Major General B.T.Surikov.

Anonymous, 1995. Assessment by the Russian Fishery Enterprises Association of the threat emanating from the CW agents dumped in the Baltic International Conference, KYOTO, Japan, 4-9 December 1995.

Anonymous, 1996. Concept of radiation decomposition of CW agents as supported by ecologists. Ministry for Environment Protection and Natural Resources, 23 February 1996, 10 pp.

Anonymous, 1996. Conclusions made by the Ecological Safety Section on usage of radiation technology to decompose CW agents. Russian Ministry document, 8 pp., provided by Major General B.T.Surikov.

Auerbach, C., 1947. The introduction by mustard gas of chromosomal instabilities in *Drosophila melanogaster*. Proc. R. Soc. Edinb. B 62, 307-320.

Auerbach, C. and Moser, H., 1950. Production of mutations by monochloro-`mustards'. Nature, 166, 1019-1920.

Auerbach, C., and Robson, J.M., 1947. The production of mutations by chemical substances. Proc. R. Soc. Edinb. B 62, 271-283.

Beale, G.H., 1995. Charlotte Auerbach. The Royal Society, Biographical Memoirs, 21-42.

Beletskaya, I.P., 1998. The Russian-US joint evaluation of the Russian two-stage process for the destruction of nerve agents. In Hart & Miller (1998), 103-112.

Brownlie, I., 1979. Principles of public international law. Claredon Press, Oxford. 3rd ed., 141 pp.

Budavari, S. (Ed.) 1989. The Merck Index. Merck & Co Inc. Rahway N.J., USA, 10100 pp. + tables.

Bundesamt, 1993. Chemische Kampfstoffmunition in der südlichen und westlichen Ostsee. Bundesamt für Seeschiffahrt und Hydrographie., 65 pp + annexes.

Chimiskyan, A., 1998. Russia on the path towards chemical demilitarisation. In Hart & Miller (1998), 14-29.

CHEMSEA, Results from the CHEMSEA project –Chemical munitions Search and Assessment. ISBN: 978-83-936609-1-9

CIRIA, 1991. Manual on the use of rock in coastal and shoreline engineering, CIRIA Special Publication 83 / CUR-Report No. 154.

CONTEX, 1995. Design developments made by the Contex Corporation on retrieval of CW ammunition from the seabed. 7 pp.

Duursma, E.K. and Carroll, L. 1996. Environmental compartments; equilibria and assessment of processes between air, water, sediments and biota. Springer-Verlag, Berlin etc. 277 pp.

Fed. Mar. Hydr. Ag., 1993. English version of Bundesamt. (1993), Chemical Munitions in the Southern Baltic Sea, Federal Maritime and Hydrographic Agency, Hamburg, 60 pp.

Federov, L.A., 1994. Chemical weapons in Russia: history, ecology, policy. Center of Ecological Policy of Russia, Moscow, 120 pp.

Federov, L.A. 1995. Chemical war in Russia: policy against ecology. Moscow.

Frondorf, M.J., 1996. Special study on the sea disposal of chemical munitions by the United States. In Kaffka (1996), 35-40.

Gerlach, S.A., 1994. Oxygen conditions improve when the salinity in the Baltic Sea decreases. Mar. Poll. Bull., 28, (7), 412-416.

Goryachev, V.L., Rutberg, F.G. and Fedyukovich, V.N., 1998. Electric-discharge method of water cleaning. State of art and prospective. Izvestiya Akademii Nauk, Energetika (in Russian), No. 1, 40-55.

Granbom, P.O., 1996. Investigation of a dumping area in the Skagerak 1992. In Kaffka (1996), 41-48.

Greenpeace, 1990. Alternative technologies for the detoxification of chemical weapons: An information Paper, prepared by Alfred Picardi (mainly focusing on general toxic waste), 91 pp. A similar document exist under the same title, but as an information Document, 104 pp.

Hart, J. and Miller, C.D. (Eds.), 1998. Chemical weapon destruction in Russia: Political, legal and technical aspects. Stockholm Int. Peace Res. Inst., 159 pp.

HELCOM CHEMU 2/2/1, 1993. Extract of the Russian report on dumped chemical weapons in the Baltic Sea. 7 pp.

HELCOM CHEMU 2/2/5., 1993. Report on sea dumping of chemical weapons by the United Kingdom in the Skagerak waters post World War II., 8 pp.

HELCOM CHEMU, 1994. Report to the 15th Meeting of the Helsinki Commission from the *ad hoc* Working group on dumped chemical munitions (HELCOM CHEMU). Danish Environmental Protection Agency, 43 pp.

HELCOM CHEMU, 1995. Final report of the *ad hoc* Working group on dumped chemical munitions (HELCOM CHEMU) to the 16th Meeting of the Helsinki Commission, 22 pp.

HELCOM 2013 Chemical Munitions Dumped in the Baltic Sea. Report of the *ad hoc* Expert Group to Update and Review the Existing Information on Dumped Chemical Munitions in the Baltic Sea (HELCOM MUNI) BSEP) No. 142 Baltic Sea Environment Proceeding pages 128.

Helsinki Convention, (1974). Convention on Marine Environmental Protection.

HMSO ,1987. Medical manual of defence against chemical agents. JSP 312. Ministry of Defence, Her Majesty's Stationary Office, ISBN 0 11 772569 2, D/Med (F&Sx2)10/11 HMSO Publication Centre, P.O.Box 276, London. SW8 5DT.

IMO, 1978. Convention on the establishment of an international fund for compensation for oil pollution damage. In force October 16, 1978. Depository: International Maritime Organisation, London. Protocol to amend in force April 8, 1981 and Protocol to amend signed May 25, 1984. Depository: International Maritime Organisation, London

IMO, 1983. Protocol relating to intervention on the high sea in cases of marine pollution by substances other than oil. In force November 2, 1983. Depository: International Maritime Organisation, London.

IMO, 1991. IMO 14th consultative meeting, 25-29/11/91, LDC 14/7/3. Long-term strategy for the convention; dumping of chemical warfare and conventional ammunition at sea, 4 pp.

IMO, 1996. Protocol to amend the International Convention on civil liability for oil pollution damage. In force May 30, 1996. (Earlier protocols, in force April 8, 1981 and signed

May 25, 1984 - did not enter in force). Depository: International Maritime Organisation, London.

Kaffka, A.V. (Ed.), 1996. Sea-dumped chemical weapons: aspects, problems and solutions. NATO ASI Series, 1. Disarmament Technologies, Vol. 7., Kluwer Academic Publishers, Dordrecht, Boston, London, 170 pp.

Kilbey, B., 1995. Obituary notice of Charlotte Auerbach. Royal Society of Edinburgh Yearbook, 1995, 87-88.

Knightley, P., 1992. Dumps of death. The Sunday Times Magazine, April 5, 1992.

Korzeniewski, K., 1994. War gases in the southern Baltic Sea. Studia I Materialy Oceanologcizne Nr. 67- Marine Chemistry (10), 91-101.

Korzeniewski, K., 1999. Chemical warfare agents dumped in the Baltic Sea, an overview. Oceanological Studies, XXVII, No. 1, 000-000 (in press).

Krutzsch, W. and Trapp, R., 1994. A commentary on the chemical weapons convention. Martinus Nijhoff Publ., Dordrecht. 543 pp.

Kurganov, R. and Morozov, V., 1999. Local ecologocal monitoring with reference to their search for sunken chemical weapons and the control. International Joint-Stock Corporation Vimpel,4 pp.

Lisichkin, G.V., 1996. Chemical weapons on the seabed. In Kaffka (1996), 121-127.

Malyshev, L.P. 1996. Technical questions of safe elimination of CW dumps on the Baltic Sea bed. In Kaffka (1996), 93-104.

Medvedeva, N., Polyak, Y., Kankaanpää, H; and Zatyana, T., 2009. Microbial responses to ùustard gas dumped in the Baltoc Sea. Mar. Env. Res., 68, 71-81.

MEDEA, 1997. Ocean dumping of chemical munitions: environmental effects in Arctic Seas. MEDEA Office, MS Z059 1820 Dolley Madison Blvd. McLean, Virg. 22012, 241 pp.

Nasrat, G.E., Kaplan, W.D. and Auerbach, C., 1954. A quantitative study of mustard gas induced chromosome breaks and rearrangements in *Drosophila melanogaster*. Z. Indukt. Abstammungs. Vererbungsl., Bd. 86, 249-262.

OPCW, 1993. Convention on the Prohibition of the Development, Production, Stockpiling and Use of Chemical Weapons and on their Destruction (OPCW Convention).

OPCW, 1997. Conference of the State Parties, C-I/DEC.8. Procedures concerning the implementation of safety requirements for activities of inspectors and inspection assistants, in accordance with part II, Par. 43, of the verification annex, 27 pp.

OPCW, 1998. Status of the Chemical Weapons Convention, S/73/98, CS-1998-831. 7 pp.

OPCW Synthesis, 1998. Newsletter of the Organisation for the Prohibition of Chemical Weapons. Issue 4, 8 pp.

OPCW, 1998. Chemical Disarmament: Basic Facts, 24 pp.

Parker, S.P. (Ed.) 1983. McGraw-Hill Encyclopedia of Chemistry, New York, 1195 pp.

Plotnikov, V.G., Zamyslov, R.A., Surikov, B.T., Dobrov, I.V. and Kayurin, O.Yu., 1996. Application of anti-filtering coatings for localisation of toxic warfare chemicals in the Baltic Sea. In Kaffka (1996), 105-107.

Rapsch, H.-J., and A. Dreher, 1996. Munition im Fischernetz Lösung eines Atlastenproblems. Atlasten Spektrum, 6/96, 1-3.

Riley, J.P. and Skirrow, G., 1965. Chemical Oceanography, Vol. I. Academic Press, London, 712 pp.

Sheluchenko, V. and Utkin, A., 1998. The role of GosNIIOKhT in the Russian chemical weapon destruction programme. In Hart & Miller (1998), 113-121.

Shmelev, V.M., Evtyukhin N.V. and Che D.O., 1996. Sterilization of water by means of pulse surface discharge. Khimicheskaya Fizika (in Russian), 15, No. 3, 140-144.

Smith, B.D., 1988. State responsibility and the marine environment: the rules of decision. Claredon Press, Oxford, 281 pp.

Soni, R., 1985. Control of marine pollution in international law. Juta & Co., Cape Town, 301 pp.

Stock, T., 1996. Sea-dumped chemical weapons and the chemical weapon convention. In Kaffka (1996), 49-66

Stock, T., 1998. Chemical weapon destruction technologies for the Russian CW stockpile. In Hart & Miller (1998), 76-93.

Surikov, B.T., 1996. How to save the Baltics from ecological disaster. In Kaffka (1996), 67-70.

Surikov, B.T., 1997. Not only the Baltic under threat. Report at the second International Conference in Italy, 1996. The Russian Social-Ecological Magazine, GREEN CROSS, No. 3, p. 2-10.

Tarasov, V.A. and Kalinina, L.M., 1992. Evaluation by Russian geneticists of the threat presented by dumped toxic agents in European seas. 12 pp. Document provided by Major General B.T.Surikov. (24 references in English.)

Theobald, N. and Rühl N.-P., 1993. Chemische Kampfstoffe in der Ostsee. Deutsche Hydrographische Zeitschrift, Supplement 1, ISSN 0946-2015, 121-131.

Theobald, N., 1994. HELCOM-Arbeitsgruppe "Chemische Kampfstoffe" - Schluszfolgerungen und künftige Aktivitäten - Deutsche Hydrographische Zeitschrift, Supplement 2, ISSN 0946-2015, 133-138

Timagenis, G.J., 1980. International Control of Marine Pollution. Oceana Publications, New York. Vol. I, 404 pp.

TNO- Prins Maurits Laboratory, R&D in weapon effects, protection and safety.

Tørnes, J.A., 1992. Investigation of ships carrying chemical ammunition sunk in Norwegian waters after World War II. Norwegian Defence Research Establishment, Div. Env. Toxicol., Kjeller, Norway, FFITO/625/138, 9 pp.

Tørnes, J.A., 1997. Investigation of ships carrying chemical ammunition sunk in Norwegian waters after World War II. Norwegian Defence Research Establishment, Div. Env. Toxicol., Kjeller, Norway, 7 pp.

Tørnes, J.A., Blanch, J.H., Wedervang, T.I., Andersen, A.G. and Opstad, A.M., 1989. Undersøkelse av skipsvrak inneholdende kjemisk ammunisjon sneket i Norske farvann eter annen verdenskrig. Cited by Bundesamt (1993).

TRUD, 1999. Russian newspaper article on CW agents. January 13, 1999.

UNCLOS III, 1982. United Nations Convention on the Law of the Sea.

Voipio, A. 1981. The Baltic Sea. Elsevier Oceanography Series, 30, 410 pp.

Volk, F., 1996. Reaction products of chemical agents by thermodynamic calculations. In Kaffka (1996), 129-143.

Yufit, S.S., Miskevich, I.V. and Shtemberg, O.N., 1996. Chemical weapons dumping and the White Sea contamination. In Kaffka (1996), 157-166.

9. ANNEXES
E.K.Duursma

9.1. Annex I: Properties of CW agents

The chemistry of CW agents is well known and details of their properties can be found in chemical handbooks. These properties determine the environmental behaviour of dissolution, hydrolysis and of their toxicity to man. In Table II some of these properties are given:116

* *Chemical formula.* A number of properties can be deduced from these formula such as: the composition of the toxic elements, their toxic configuration and their possible breakdown products after hydrolysis in sea water; for example, the amount of arsenic liberated after ultimate decomposition.

* *ICT50.* This abbreviation is the incapacity concentration/time, calculated from the exposure time in air at which 50% of contaminated men would be incapacitated (the exposure is given in minutes and multiplied by the amount in the air (mg/m3)). At low concentrations (mg/m3) this will be longer than at higher concentrations.

* *LCT50.* This abbreviation is the lethal concentration/time, calculated from the exposure in air at which 50% of the contaminated men would die, also depending on a shorter or longer exposure. For example 50% would die after one minute inhaling 380 mg tabun/m3 or after ten minutes at 38 mg/m3.

* *Volatility.* The volatility is the maximum amount of toxin in air when exposed in a closed space. The values differ greatly and should be compared with the ICT50 and LCT50 values, given above. For example, the volatility of tabun exceeds that of its ICT50 and LCT50 values (intoxication by inhalation is relatively rapid), whereas the volatility of S-Mustard is lower than its ICT50 and LCT50 values (intoxication by inhalation is relatively slow).

* *Melting Point.* The melting points are usually below freezing point, except for S-mustard (+14oC), and the arsenic toxins adamsite (195oC), clark I (44oC) and clark II (33oC).

* *Boiling Point.* The boiling points are high (above 100oC) except for phosgene (8.2o) and cyanic acid (23.5oC). This means that at the local temperatures in the Baltic Sea and Skagerak, which, at the sea bottom are usually below 15o, these toxins behave as solids (adamsite, clark I and clark II) or as liquids, except S-mustard which is a solid product below 14oC. Liquid mustard is converted into leathery lumps, sticking to sedimentary material.

* *Solubility.* The solubility of the toxins in sea water depends on their so-called polarity. When their polarity is low, they do not dissolve easily in water, which is a polar liquid. Therefore N-mustard, S-mustard and lewisite have low solubility in water. They are much more soluble in non-polar liquids, such as petrol gasoline and lipids (fat).117

* *Density.* Except for hydrocyanic acid, the CW agents in liquid form have a higher density than that of sea water of 0-34 ☐ Salinity (1.00-1.027 g/cm3).

* *Log KOW.* This term is the logarithm of the ratio (K) between the solubility of a substance in octane (O) and its solubility in water (W). When log KOW = 1, 2 or 3, the ratio KOW = 10, 100 and 1000, respectively. Therefore the higher KOW value for mustard and lewisite coincide with their relatively low solubility in water and high affinity for lipids.

116Bundesamt (1993), Fed. Mar. Hydr. Ag. (1993), HELCOM (1994), Volk (1996) and MEDEA (1997).
117This can be taken into account when decontaminating ship-decks and the skin of victims.

Table IIA. Properties of CW agents (Cl-acetophenone as tear gas is not included): Chemical formula, Incapacity concentration/time, Lethal concentration/time and Volatility. For references see: Bundesamt (1993), Fed. Mar. Hyd. Ag. (1993), HELCOM (1994), Volk (1996) and Parker (1983).

Type of CW gasses	Chemical Formula	ICT_{50} min.x.mg/m^3	LCT_{50} min.x.mg/m^3	Volatility mg/m^3
Nerve agents (Organo-phosphorus compounds) Tabun, GA Sarin, GB Soman, GD	Dimethylphosphoramidecyanide acid, ethyl ester Methylphosphonofluoride acid, (1-methylethyl) ester Methylphosphonofluoridic acid, 1,2,2-trimethylpropyl ester	10 (inhalation)	380 (inhalation)	610 (25°) 21,900 (25°) 3060 (25°)
Blood agents Hydrocyanic acid, AC, Prussic acid Cyanogen-chlorine, CK	Hydrocyanic acid Cyanogen chloride		(Lethal Dose) 0.57 mg/kg body weight	Large
Lung-damaging (choking) agents Phosgene, CG Diphosgene, DP	Carbonic dichloride Carbonchloride acid, trichloromethyl ester	1600 1600	3200 3200	3,200,000 (0°)
Vesicant (blister) agents N-Mustard (HN N-Lost) S-Mustard (Yperite, H, S-Lost) Lewisite, L	2-Chloro-N-(2-chloroethyl)-N-ethylethanamine 1,1'-Thiobis(2-chloroethane) (2-Chloroethenyl) arsonous dichloride	2000 (skin) 1500 (skin)	10,000 (skin) 100,000 (skin)	1290 (20°) 628 (20°) 4480 (20°
Nose and throat irritant agents Adamsite, DM Clark I, DA Clark II, DC	10-Chloro-5, 10-dihydrophenarsazine Diphenylarsinous chloride Diphenylarsinous cyanide	8 20 12	15,000 15,000 10,000	0.02(20°) 6.8 (20°) 2.8 (20°)

Table IIB. Properties of CW agents: Melting point, Boiling point, Solubility, Density (Specific weight) and Log K_{OW}. For references see: Bundesamt (1993), Fed. Mar. Hyd. Ag. (1993), HELCOM (1994) ,Volk (1996), MEDEA (1997) and verified on Budavari (1989) = Merck Index.

Type of CW gasses	Melting Point ^{o}C	Boiling Point ^{o}C	Solubility g/l	Density g/cm^3	Log K_{OW}
Nerve agents (Organo-phosphorus compounds)					
Tabun, GA	-50	240	72-120	1.07	-1.44
Sarin, GB	-57	147	miscible	1.11	0.31
Soman, GD	-80	167			
Blood agents					
Hydrocyanic acid, AC, Prussic acid	-86	23.5	100%	0.7	-0.69
Cyanogen-chlorine, CK					
Long-damaging (choking) agents					
Phosgene, CG	-118	8.2	9	3.4	
Diphosgene, DP	-57	127		1.65	
Vesicant (blister) agents					
N-Mustard (HN N-Lost)	-34	235	0.16	1.24	
S-Mustard (Yperite, H, S-Lost)	14	217	0.8	1.27	1.4-2.4
Lewisite, L	0	190	0.5	1.89	2.3
Nose and throat irritant agents					
Adamsite, DM	195	410	0.002	1.65	
Clark I, DA	44	333	2	1.42	
Clark II, DC	33	346	2	1.45	

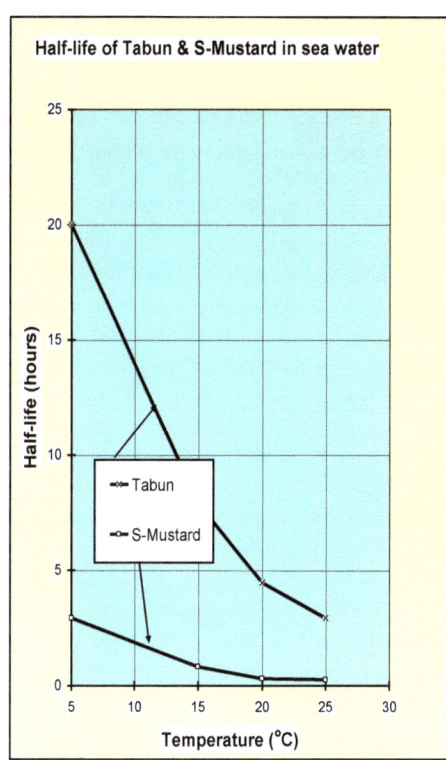

Fig. 43. Half-lives of tabun and S-mustard for a temperature range from 5 to 25o C.

Fig. 44. Disappearance percentage of a dissolved CW agents in time elapsed of half-lives.

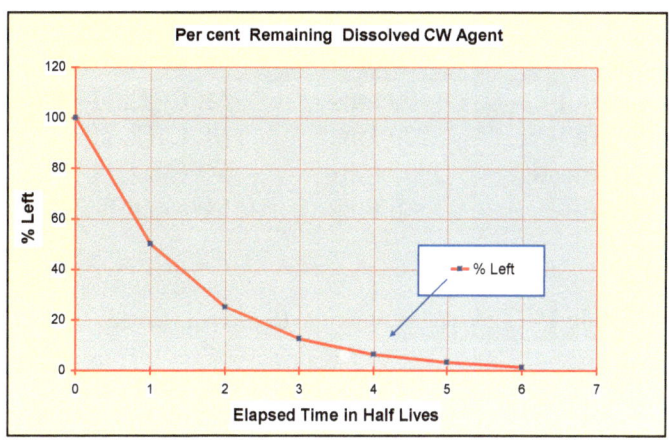

Hydrolysis

The hydrolysis of four CW agents is well described by the article MEDEA (1997).118 The conclusions are:

* Tabun is fairly soluble in water and hydrolyses over a period of hours (Figs. 35 and 36).
* Sarin is miscible (mixes in all proportions) with water and also hydrolyses over a period of days.
* Dissolved S-mustard hydrolyses relatively rapidly (**Figs. 42 and 43**). However, the persistence of mustard in the marine environment is controlled by the rate at which it dissolves and this is very slow. Thus few hydrolysis products can be expected.
* Lewisite is soluble in water and hydrolyses very rapidly. The initial hydrolysis products are also very toxic and persist in sea water for months or longer before finally being converted to arsenic.

General

Hydrolysis is usually more rapid under alkaline conditions and it may be worthwhile creating alkaline conditions when confining CW agents and prior to treatment.

---

[118]MEDEA (1997), 4. Sea water chemistry, 4-1 - 4-26.

9.2. Annex II: Calculation of possible arsenic contamination of the Baltic Sea

Although no measurable arsenic contamination has been determined in the Western Baltic Sea,[119] the maximum levels, following a liberation of all arsenic - supposing all German produced arsenic CW's were dumped in the Baltic Sea -, can be calculated from the data of HELCOM CHEMU (1994).[120]

Table III: Calculation of molecular weights.

Name	Amounts	Molecular weight (=M)		
Clark I	1500 tons	M: C-12, H-10, As, Cl:	144+10+75+35.5	= 264.5
Clark II	100 tons	M: C-12, H-10, As, N:	144+10+75+14	= 243.-
Adamsite	3900 tons	M: C-12, H-9, As, N, Cl:	144+9+75+14+35.5	= 277.5
As-oil	7500 tons	M: roughly 200.		

Table IV: Calculation of weight fraction of arsenic.

Name	As fraction (75/M)x100%	Tons of As
Clark I	28.4%	426.- tons
Clark II	30.9%	30.9 tons
Adamsite	17.0%	1053.- tons
As-oil	37.5%	2812.5 tons
		4322.4 tons

Considering the mean natural arsenic content of Baltic seawater to be 1 μg/l, (= 1 mg/m3 = 10-3 g/m3 = 10-9 tons/m3), a water column of 4322.4 x 109 m3 (or 4322.4 km3) would be required to increase the arsenic content by 1 μg/l.

Since the total volume[121] of the Baltic Sea is 21,000 km3, complete mixing would only increase the arsenic concentration by 20%.

Considering the residence time of this water is about 27-36 years, leakage of arsenic toxins over long periods of several decades or centuries would not contaminate the Baltic Sea with arsenic, other than locally around the dump sites.

---

[119]Theobald and Rühl (1993), p. 128.
[120]HELCOM CHEMU (1994), TABLE 1, p.10.
[121]Voipio, 1981, p. 352.

9.3. Annex III: Positions of localised wrecked ships loaded with CW agents in the Skagerak

The degree of corrosion reached by the CW cargoes in the sunken ships in the Skagerak should be solved. The USA operation called *Davey Jones Locker* sunk the following vessels (the last two in the Northern North Sea):

Table V: Ships sunk during the USA operation *Davey Jones Locker* (1946-1948).

Name ship	L. Tons	Depth (m)	Lat./Long.
Sperrbecher (Mine breaker)	1349	650	58o14'N/9o15'E
T-65 (Flak ship)	1526	650	58o17'9"N/9o37'1"E
U.-J. 305 (Trawler)	671	650	58o16'4"N/9o29'E
Alco Banner (?)	2765	650	58o18'7"N/9o36'5"E
James Otis (Liberty)	3653	680	58o16'N/9o32'E
James Sewell (Liberty)	4000	725	58o15'2"N/9o30'6"E
James Harrod (Liberty)	3000	665	58o16'0"N/9o33'0"E
George Hawley (Liberty)	1000	685	58o18'5"N/9o38'0"E
Nesbitt (Liberty)	6000	580	58o18'5"N/9o30'0"E
Philip Heiniken (Freighter)	2000	1035	62o57'0"N/1o32'0"E
Marcy (Freighter)	2500	1180	62o59'0"N/1o23'0"E

None of these ships were identified by name during investigations carried out by the Norwegian Defence Research Establishment[122] in 1989. In a first attempt, the wrecks were located by side-scan sonar (Table VI), and five of the wrecks were selected for investigation by an unmanned remote-control vehicle.

Table VI: Position of localised ships wrecked in the Skagerak.

Wreck/ site no.	Latitude 58o **'**.*"N	Longitude 9o **'**.*"E	Possible Length; NAME
(Note 1)1	13'56.5"	32'09.7"	100m; Possibly DUBORG
2	15'05.8"	34'31.0"	70m; Possibly PATAGONIA
3	15'20.9"	43'21.1"	80-100m
4	15'45.5"	42'11.5"	?
5	15'53.6"	40'01.9"	110m
6	16'08.5"	41'08.6"	Broken in several parts
7	16'00.9"	31'15.2"	70-80m; Possibly TAURUS
8	17'42.0"	40'42.0"	100m
9	17'52.7"	41'34.0"	100m
10	17'52.8"	38'46.0"	70m
11	17'28.8"	33'08.5"	130m
12	17'30.2"	25'11.7"	?
13	18'47.0"	39'56.2"	120
14	18'31.5"	41'05.7"	100m; Identified SESOSTRIS
15	19'44.9"	39'43.6"	?

One of the wrecks had broken into several parts and aircraft bombs and cases were spread out nearby. Most of the bombs were intact, but some had corrosion holes (see Fig. 12). Water samples were taken from the hole of a corroded bomb and above loading doors of other wrecks. Mustard gas and nerve agents were not found above the detection limit.

---

[122]Tørnes (1997).

9.4. Annex IV: Nautical charts of discussed dump sites

Official nautical charts, some of them indicating the areas where fishing and/or anchoring is forbidden. The copies of the charts were obtained by the courtesy of the International Hydrographic Bureau in Monaco. The HELCOM areas (see Fig. 1) where it is recommended to forbid fishing are added as dashed squares. Chart A. Baltic Sea south of Gotland.

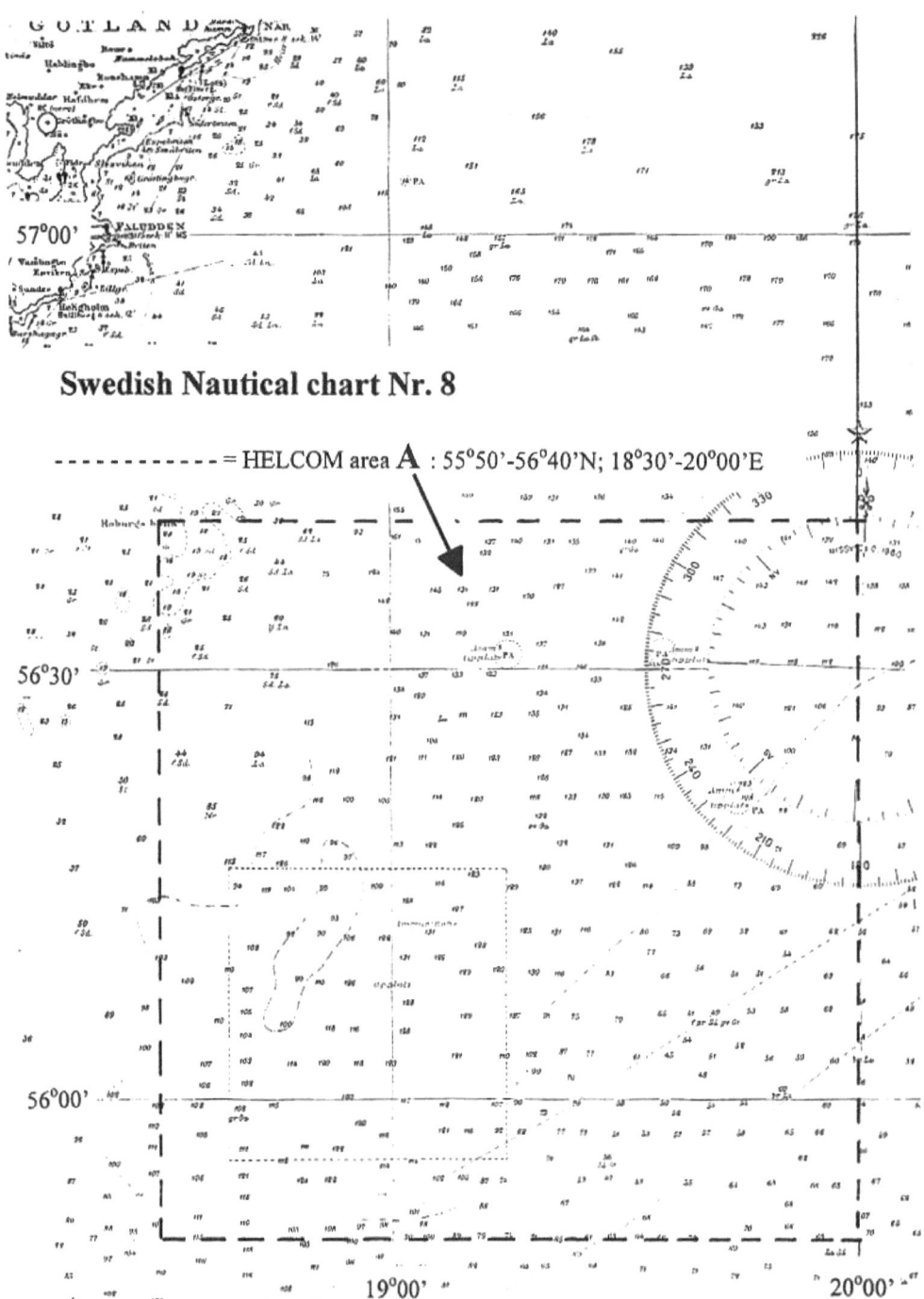

Swedish Nautical chart Nr. 8

- - - - - - - - - - = HELCOM area **A** : 55°50'-56°40'N; 18°30'-20°00'E

Chart B. Area around Bornholm

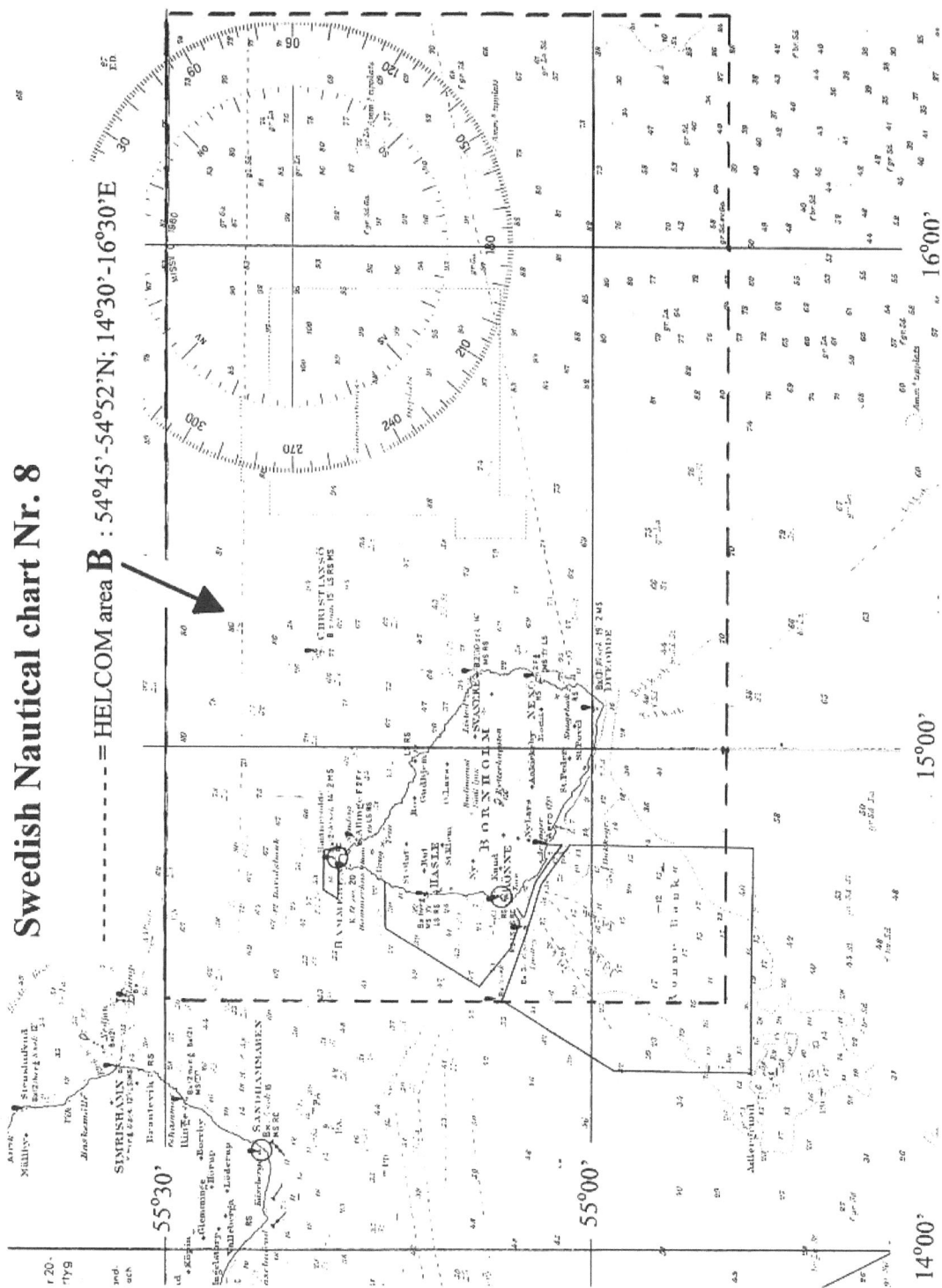

Chart C. Little Belt.

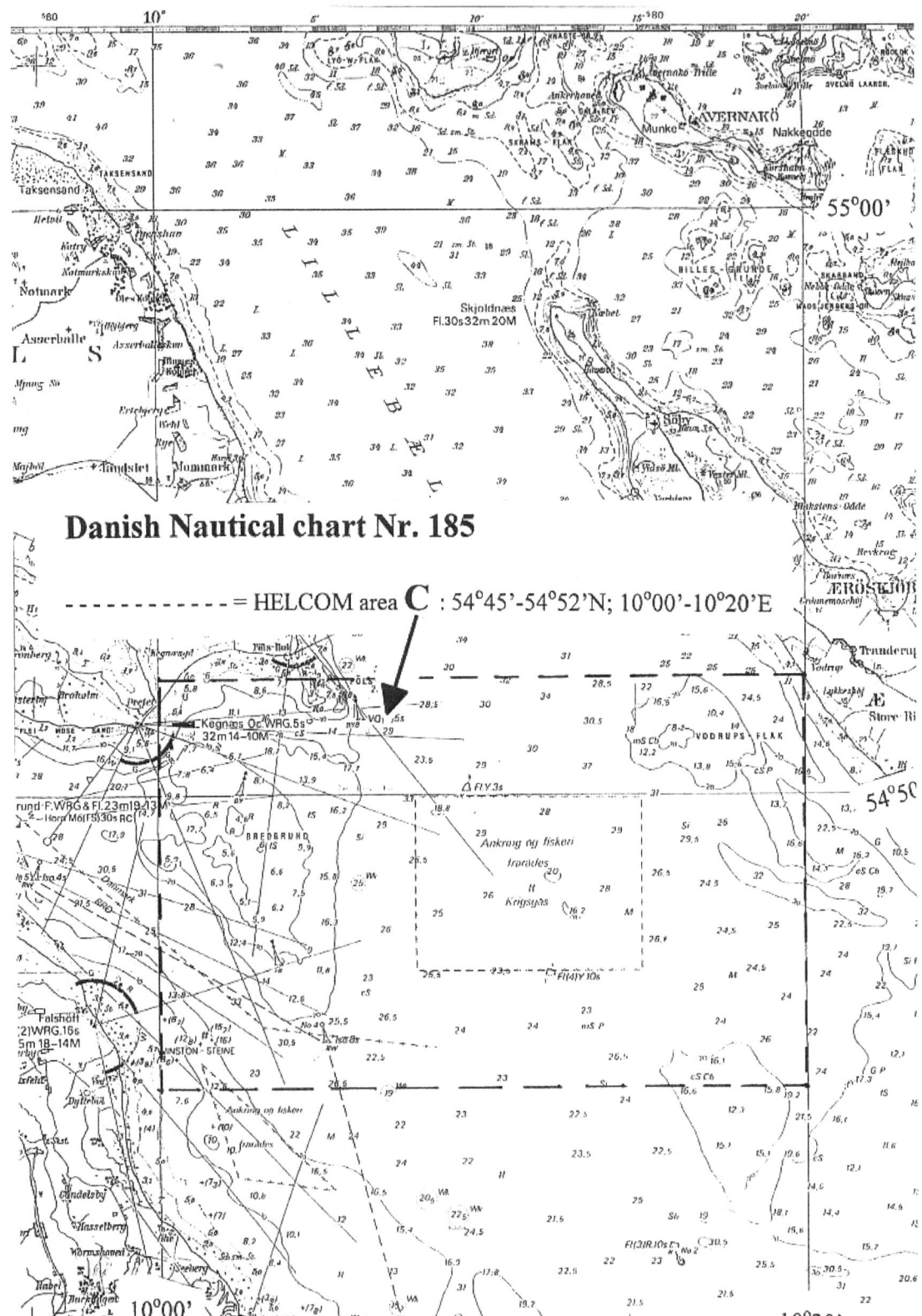

Danish Nautical chart Nr. 185

- - - - - - - - - = HELCOM area **C** : 54°45'-54°52'N; 10°00'-10°20'E

Chart D & E. Skagerak. The numbers in D concern the positions of the localised wrecked ships given in Table VI of Annex III.

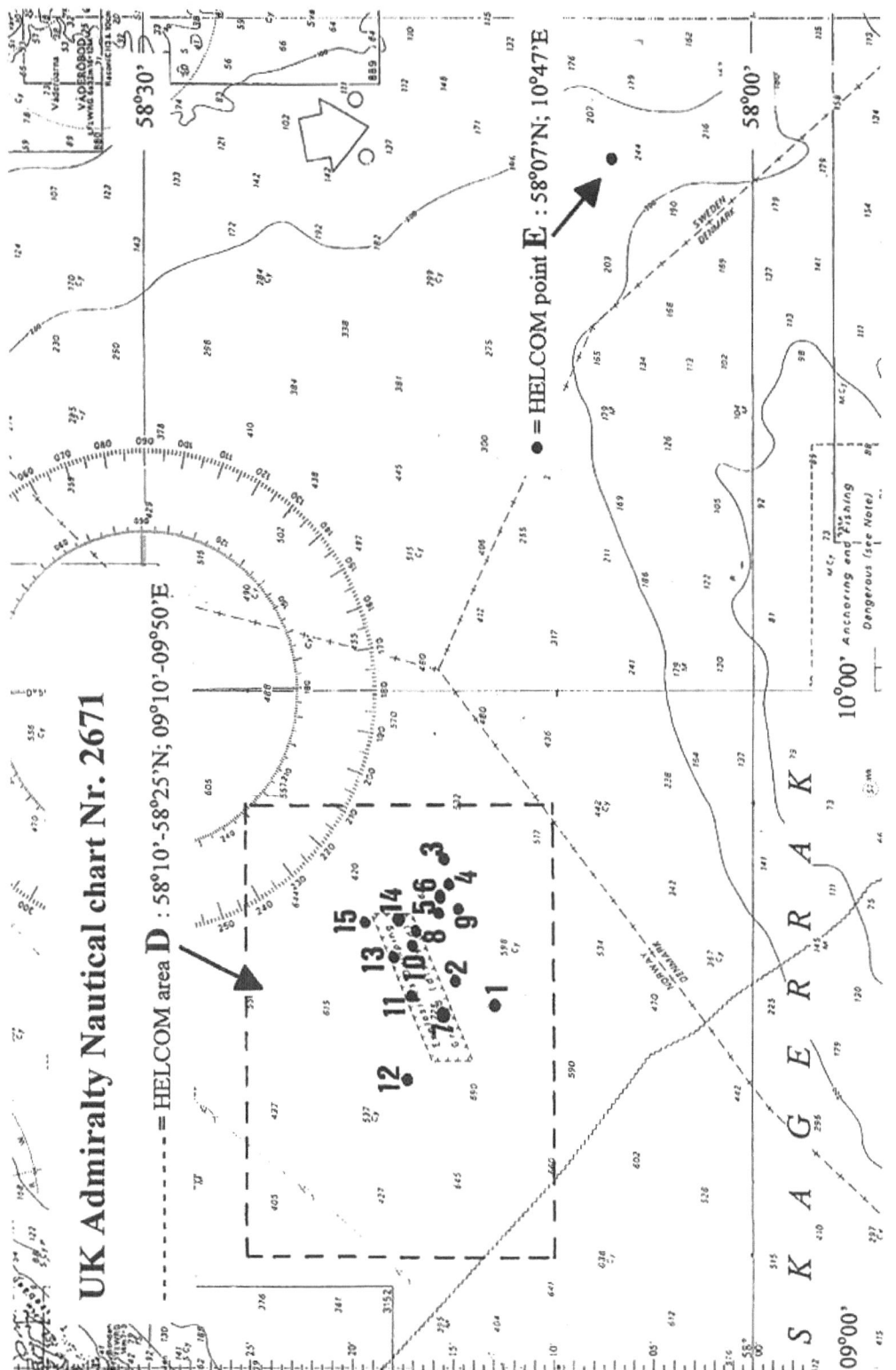

9.5. Annex V: Authors[123]

Major General Boris T. Surikov,

Retired Airforce Commander, Moscow, Russia

Candidate Science & Technology

Advisor Soviet-US Salt-1 in drafting the Treaty on the Limitation of Anti-Ballistic Missile Systems.

Expert to Russian Parliament on problems of nuclear armament reduction

Ir. Jan G. de Vries

Retired Engineering Specialist, TIDEWAY b.v., Breda, Weth. Crezeelaan 9, 4927 AC Hooge Zwaluwe, The Netherlands

Prof. dr. Igor A. Kossyi,

Institute of General Physics of the Russian Academy of Sciences, Moscow, Russia

Ir. Alexander I. Mikulin,

President Company Ecotransenergomash, Moscow, Russia

Drs. Jan H. Holsboer,

Board Member of the ING GROUP (Banking and Insurance), Amsterdam, The Netherlands

Mr. Robert C. Seward,

Britannia Steam Ship Insurance Association, London, U.K.

Dr. Jorri C. Duursma,

Laan van Meerdervoort 163, 2517 AZ The Hague, The Netherlands

Scholar in International Public Law,

Dr. Alfred H. Heineken,

Pentagon, Amsterdam, The Netherlands

Retired PDG Heineken Breweries

Doctor Honores Causa (1989, Rochester University and 1996, Hofstra University, both New York State, USA).

Prof. dr. Egbert K. Duursma (and editor)

302 av du Semaphore, 06190 Roquebrune Cap Martin, France

Retired director Netherlands Institute of Sea Research (NIOZ) and emeritus Professor in Oceaology of the University of Groningen.

Member Academia Europaea

9.6. Annex VI: Acknowledgements

Of the editor

The preparation of this synopsis would not have been possible without the help of the library of the Musée Océanographique in Monaco and of those, specified below who supplied valuable documents on CW agents and CW ammunition. As editor I am very grateful to Dr. JoLynn Carroll, Tromsø, Norway for sending me a copy of the MEDEA report and some Norwegian reports, to Mrs. Nadine McNeil and Dr. Ralf Trapp of the OPCW in The Hague, Netherlands for providing a great number of relevant documents and copies of Russian articles, to Dr. Manfred K. Nauke of IMO in London for giving the opinion of the IMO, to Dr. Klaus Löwe of the Bundesministerium für Umwelt Naturschutz und Reaktorsicherheit in Berlin, to Dr. Norbert Theobald of the Bundesamt für Seeschifffahrt und Hydrographie in Hamburg and to Dr. Hans-Jürgen Rapsch of the Niedersächsisches Umweltministerium in Hannover for sending me all German documents on CW agents dumped in the Baltic Sea.

The help on defining the problems and indicating the NATO documents by Dr. M. van Zalm, Programme director of TNO in Rijswijk, The Netherlands, was greatly appreciated. Valuable information was also obtained from Mrs. Nancy Schulte, Programme Director Disarmament Technology of NATO in Brussels, Belgium. Ir. George F.M. Remery and Ir. Paul van Berkel, Both of Single Buoy Moorings, Monaco, and Mr. Klaas Reinigert, Vlaardingen, The Netherlands, salvage expert and former director Smit-Tak gave valuable advice and suggestions.

The secretary of the Helsinki Commission, Dr Joni Kaitaranta and Dr. Kjell Grib, Finland, and Mr. Kjeld F. Jørgensen, Denmark, provided all requested documents of the *ad hoc* Working Group HELCOM CHEMU, to which reference is made in this synopsis. The International Hydrographic Bureau (IHB) in Monaco, by its President Rear Admiral Giuseppe Angrisano, made a request for additional information to the hydrographical

---

[123] Of first edition (1999) and cited in this document

bureau's of its Member States for which I am greatly indebted. The IHB, by Mr. René DelFa, also provided copies of the recent nautical charts. They are presented in Annex IV.

Most details on symptoms and first-aid as given in Table I were obtained from Deputy Chief, Commandant Christian Chevallier and Adjudant Jean-Marc Decaunes of the Sapeurs-Pompiers de Monaco and through Rear-Admiral Gilbert de Cock from a document prepared by the UK Her Majesties Stationary Office. These data were verified by Dr. René D'Agro, dermatologist in Menton France and with information available in the obtained German and HELCOM documents. The Direction North Sea of the Rijkswaterstaat, Rijswijk, The Netherlands gave valuable information about their rules on handling the financial compensation on captured ammunition in the North Sea by fishermen. Prof. Geoffry Beale of the University of Edinburgh is greatly acknowledged for his very prompt replies and documents on the late Dr. Charlotte Auerbach expert in genetic effects of mustard gas. Mr. John Aasulf Tørnes of the Norwegian Defence Research Establishment is thanked for providing valuable literature and the permission to reproduce photographs of corroded bombs in the Skagerak. Dr. Krzysztof Korzeniewski, Institute of Oceanography, Gdynia, Poland kindly gave permission to reproduce a figure from his publication on War gases in the southern Baltic Sea. Mrs Jasmine Antonini-Kane is credited for correcting the English text.

Naturally all participating authors are acknowledged for writing essential chapters in this synopsis and supporting my editorial comments and arrangements.

A number of illustrations were reproduced from the video-film Cain's Smoke, for which kind permission was obtained. The producer's address is: Creative Ass. ⬜Film Program - the XXth Century⬜, Mr. R. A Konbrant, 8, Ezenshtein Str., A. M. Gorki Film Studio, Moscow, 129336, Tel.: 00-7-095-181-6518; Fax.: 00-7-095-181-6006.